Birding with AI

Data in the Wild is a series of practical books, sharing key tools and methods for the collection, analysis and interpretation of environmental data. They are published rapidly in print and digital formats.

https://pelagicpublishing.com/collections/data-in-the-wild

Birding with AI

Concepts and Projects for Ornithology

Ronald T. Kneusel, Ph.D.

DATA IN THE WILD SERIES

Pelagic Publishing

First published in 2025 by
Pelagic Publishing
20–22 Wenlock Road
London N1 7GU, UK

www.pelagicpublishing.com

Birding with AI: Concepts and Projects for Ornithology

https://doi.org/10.53061/RIEY9242

A CIP record for this book is available from the British Library

ISBN 978-1-78427-602-7 Hbk
ISBN 978-1-78427-603-4 ePub
ISBN 978-1-78427-604-1 PDF

EU Authorised Representative: Easy Access System Europe – Mustamae tee 50, 10621 Tallinn, Estonia, gpsr.requests@easproject.com

Cover images: Green-crowned Brilliant Hummingbird in flight La Paz Waterfall Garden, Costa Rica: Unsplash/Zdeněk Macháček. Circuit board, iStock/Ariq stock.

Typeset in Palatino LT Std by S4Carlisle Publishing Services, Chennai, India

Contents

Figures

Introduction

The recent emergence of powerful *artificial intelligence (AI)* systems has opened new avenues for birding and ornithology, avenues worthy of exploration and discussion; hence this book.

AI has permeated modern life and isn't through with us yet. Anyone with access to a smartphone has all they need to participate in birding and, more importantly, citizen science. AI-powered applications, like Merlin from Cornell University, instantly identify birds by sound or image. Gone forever are the days when a would-be birder must memorize a plethora of characteristics to distinguish one sparrow from another, or to differentiate between the call of a Red-winged Blackbird and a Common Grackle. There remains merit in learning such, of course, but AI has opened the door to many who would never have taken the time otherwise. Not surprisingly, time spent with such tools builds that knowledge unconsciously.

For ornithologists, AI is rapidly becoming a necessary research tool. With AI, it becomes possible to classify camera trap images or any other image source automatically. Likewise, research tools like BirdNet, also courtesy of Cornell University, permit bulk interpretation of large collections of recordings, thereby eliminating the drudgery of listening carefully to hours of bird song to mark which species were present and when. And among graduate students the world over, there was much rejoicing.

This book introduces ornithologists and the birding community to AI via a three-part narrative: (1) what AI is and how it works, (2) gaining experience in the design and implementation of AI research projects by exploring hands-on mini-projects and (3) AI in birding using existing, often free, tools.

Artificial intelligence is a subfield of computer science that began in the 1950s. One aspect of artificial intelligence, known classically as *connectionism*, attempts to develop intelligent machine behavior from the ground up using collections of cleverly arranged and trained basic units, usually denoted as *neurons* for their superficial similarity to biological neurons. I prefer the term *nodes*, which is correct from a computer science perspective, but I'll use both terms interchangeably throughout the book.

Historically, connectionism was a poor cousin to the other aspect of AI, *symbolic AI*, which attempted to encode intelligence symbolically; think logical statements expressed mathematically and inference over relationships.

A more complete history of the two approaches, both present in the 1950s, is found in my book *How AI Works* (No Starch Press, 2023), but at the risk of oversimplifying: decades of work in symbolic AI achieved only (very) modest success, and connectionism, at times actively suppressed, exploded in the early 2010s with the advent of *deep learning*, a catch-all term referring to large *neural networks* with many nodes and layers. The AI revolution prompting this book is all deep learning, all neural networks. In short, the connectionists won the day, and when people now speak of "AI," they invariably mean deep learning with neural networks.

The book's first part, Chapters 1 and 2, describes how AI works. The focus is on understanding AI concepts and commonly used terms to allow you to follow AI discussions and understand the general workflow of data → model → training → inference. Building AI systems requires data from which models are constructed and trained to be used in the wild to make inferences.

The final part, Chapter 10, introduces currently popular AI-based birding tools for classifying images and audio. Discussing such tools gives us an opportunity to put our newfound AI knowledge into practice. AI isn't perfect, so understanding how it works allows you to interpret the tools' behavior critically.

The second part, Chapters 3 through 9, is where we get our hands dirty. We explore AI by doing AI. Naturally, birds are our target, but you'll no doubt notice that other datasets can replace the bird datasets. In the end, neural networks are unaware of the meaning of their inputs. It's all numbers to them, so replacing bird images with mammal or insect images won't alter the process and should still produce valuable results. Consider the projects as exercises pointing the way to more elaborate research projects. If you finish the book with a new project idea to address one of your research questions, I will have succeeded.

Lastly, you likely noticed several terms in the preceding paragraphs are *emphasized*. Such terms appear in the glossary at the end of the book. I hope you find it useful.

****

The poet John Donne wrote, "No man is an island." In that vein, no book is an island, even if written by a single individual. I'd like to thank those who assisted and made this book a reality.

First, Nigel Massen at Pelagic Publishing, who suggested the book and then made it possible.

Next, my family, for putting up, yet again, with hours spent cloistered in my office, to say nothing of the many hours spent wandering through woods and grasslands and by lakes and streams to gather the photographs and audio found in this book. It's just another one of Dad's pet projects.

****

You'll find all the book's source code on GitHub:

```
https://github.com/rkneuse19/BirdingWithAI/
```

with relevant datasets downloadable from the GitHub page.

Questions or comments? Contact me: rkneuselbooks@gmail.com

"Wherever there are birds, there is hope."
  – Mehmet Murat ildan

1. AI in a Nutshell

Using AI effectively implies an understanding of what it is, where it came from, and, most importantly, how to apply it to our data.

In this chapter, we learn what "artificial intelligence" refers to and how the term relates to other terms like "machine learning" and "deep learning," which you'll encounter at various points in your AI journey. Next comes a (very) brief history of AI to place it in context and to clarify why now, after decades of false promises, AI has truly arrived.

The following two sections discuss neural networks, the backbone of modern AI, and the datasets used to train and test such networks. The book's projects use neural networks, datasets, and training and testing, so we must comprehend their relative contributions to the process. The dataset section presents our first neural network model, which I invite you to review carefully.

Be forewarned: there be mathematics ahead. It's not a lot of mathematics, however, just enough to gain the level of understanding we need, but don't say I didn't warn you.

1.1 Defining AI

Artificial intelligence is the subfield of computer science that attempts to evoke "intelligent" behavior from computers. That's a nebulous definition, to be sure, but adequate. It's up to us to decide what qualifies as intelligent. The long-term goal of AI is to create machines that match or exceed human intelligence and intellectual ability. Fortunately, our goals are considerably more modest – all we desire is to build systems that identify birds in images and audio.

To a computer scientist, AI refers to a rich set of algorithms and approaches reaching back to the early 1950s. However, when people mention "AI" colloquially, they are invariably referring to what is known as *deep learning (DL)*, which is itself a subfield of *machine learning (ML)*, a subfield of artificial intelligence proper. I think of the relationships graphically:

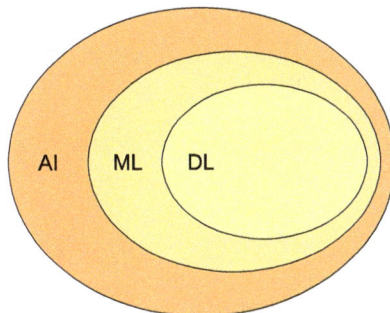

Machine learning is the part of AI that uses computers to recognize patterns in data. Neural networks are just one of many machine learning algorithms. Deep learning uses neural networks, specifically large, multilayered neural networks, to accept inputs (think images) and produce useful outputs (think "it's a Blue Jay"). Modern AI is at present all neural networks, though this wasn't always the case.

1.2   A Brief History of AI

The history of AI is long, rich and well worth investigating at some point. It suffices for us to review only the most salient events, if for no other reason than to understand why, with over 70 years of effort behind it, AI is only now coming to fruition.

Most people agree that artificial intelligence as a serious research venture began with the Dartmouth Summer Research Project on Artificial Intelligence workshop of 1956. It was there that term "artificial intelligence" was first consistently used.

I separate AI into two main camps: *symbolic AI* and *connectionism*. Symbolic AI seeks to emulate intelligent behavior via logical processing, meaning the manipulation of logical statements and rules divorced from any reliance on a physical substrate, either actual or simulated. Think of symbolic AI as a top-down approach using algorithms and rules to manipulate symbols and statements to achieve intelligent outcomes.

Connectionism, on the other hand, is bottom-up. Connectionism is inspired by living brains, massively complex conglomerations of interconnected basic units, i.e., neurons. The vast array of animal life on Earth proves that this "if you build it, they will come" approach leads somewhere useful.

Both forms of AI were present at the beginning. Most researchers focused on symbolic AI, which was a reasonable thing to do given the primitive nature of computer systems in the 1950s and onward until the early 2000s. However, a few brave souls adopted connectionist ideas, including Frank Rosenblatt, who in the late 1950s developed the *Perceptron*, a hulking machine implementing what we now recognize as a neural network.

The Perceptron's abilities were modest, at best, but it did work and led to bold predictions on Rosenblatt's part about what would be coming in only a few years. Rosenblatt promised a machine that would "walk, talk, see, write, reproduce itself and be conscious of its existence" with later Perceptrons "able to recognize people and call out their names and instantly translate speech in one language to speech and writing in another language" as reported by *New York Times* in 1958. Clearly, Rosenblatt was overly optimistic, but consider: virtually everything he claimed connectionism to be someday capable of has now come to pass (save consciousness).

The intrigue found in a detailed history of AI would make for a good movie, but all that can be said here is that symbolic AI, with its modest-at-best successes, dominated for decades, at times even actively suppressing connectionist impulses (and the careers of those with them).

Everything changed around 2010 when Fortuna smiled on humanity with the confluence of three essential items that made connectionist dreams a reality: data, speed and algorithmic improvements.

Neural networks learn from data, as we'll spend most of our time in this book demonstrating in great detail. To that end, large datasets are required to train neural networks. The advent of the World Wide Web provided the impetus for an explosion of online data.

Training neural networks requires vast amounts of compute. Training large neural networks, for instance *large language models (LLMs)* such as ChatGPT or Claude, requires astronomical amounts of compute. Computers of the twentieth century were simply not up to the task. What changed around 2010 was the realization that the *graphics processing units (GPUs)* initially developed for video games were amenable to the kinds of calculations necessary to train and implement neural networks. Think of GPUs as special-purpose supercomputers tailored to the type of calculations neural networks need.

The final piece of the deep learning revolution involved improvements to algorithms. The concepts behind what became neural networks were first articulated in the 1940s. Rosenblatt's Perceptron embodied these ideas, and they persisted and grew slowly over the decades, maturing in the late 1990s and early 2000s, thanks to the work of researchers like Geoffrey Hinton, Yann LeCun, and Yoshua Bengio, among many others. The most important of the algorithmic improvements concerns how to train a neural network and how to initialize it in the first place.

Data, processing speed, and algorithms enabled the construction of large neural networks big enough to do what practitioners expected all along, leading quickly to success after success from efficiently classifying the content of digital images (hooray, we say!) to the promise of an AI-enabled, or at least AI-enhanced, future.

Neural networks are the heart of all modern AI systems. Fortunately, all neural networks, from the small ones we'll use in this book to impressive ChatGPT-style LLMs, are based on collections of creatively arranged basic units, a unit so basic it's a marvel that so much can come from so little.

1.3   Neural Networks

Take a moment to contemplate Figure 1.1. It's the foundation of every machine learning model we'll explore in this book. Consequently, it's worth our time to understand it in detail – not that there's a lot of detail to understand.

Let's move from left to right. The figure shows three squares labeled x_0, x_1 and x_2. These represent inputs to the neuron. In practice, they are real numbers like 4, 1.414, 255 and so on. The arrows indicate that the inputs move toward the circle, representing a network node. The h label represents the *activation function*, the mathematical operation performed by the neuron. The input arrows are labeled: w_0, w_1 and w_2. These are *weights*. Training a neural network involves finding the correct weights such that training inputs produce

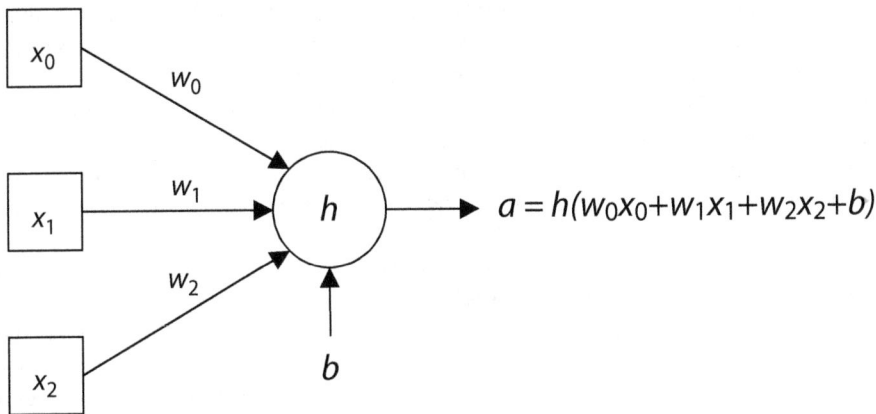

$$a = h(w_0 x_0 + w_1 x_1 + w_2 x_2 + b)$$

Figure 1.1 The humble neuron

expected output values. There is an additional arrow labeled b. This is the *bias*. Like the weights, the bias is learned during training and represents a constant offset applied to the argument of the node's activation function. The node itself multiples the input xs by their corresponding weights, w, sums the resulting products along with the bias, and passes that value to the activation function, h, to produce the output, a. Mathematically,

$$a = h(x_0 w_0 + x_1 w_1 + x_2 w_2 + b)$$

giving a from the inputs using the weights and bias learned during training. And, that is just about all there is to the operation of most neural networks. The remainder is arranging neurons in different configurations, with additional types of operations I'll introduce as needed, followed by training to assign values to the weights and biases.

The superficial resemblance between Figure 1.1 and a biological neuron motivates the name "neural network." Biological neurons accept multiple inputs (the xs in the figure) and produce an output based on some function using the inputs. The analogy, however, is weak. The neuron of Figure 1.1 is continuous, it produces a continuously varying output based on continuously varying inputs. Biological neurons are all or nothing: either the inputs exceed some learned threshold, resulting in an output, or the neuron remains inactive.

It's fair to arrange nodes as desired, but specific patterns have shown themselves to be helpful in practice. One is the classical approach of stacking layers of nodes to produce a fully connected *multilayer perceptron (MLP)*, a sequence where every input to a layer is directed to every node of the layer, with the output of the layer's nodes the input to the next layer. The other, the one most relevant to us, is the *convolutional neural network (CNN)*, the architecture used to interpret images. Let's use the MLP's architecture as a stepping stone to the CNNs. Once we have CNNs, we're in business.

Consider the remainder of this section to be background material. It is entirely possible (though I discourage it) to work through the projects in the book without understanding what's happening behind the scenes. Fight the temptation to skip ahead. The effort spent will reap rewards later, not just for the projects but for navigating an increasingly AI-dominated world.

1.3.1 Multilayer Perceptrons

We begin with MLPs to understand essential terminology and gain helpful intuition about what neural networks actually do. Everything discussed in this section applies directly to convolutional networks, only the level of complexity changes.

Figure 1.2 is an MLP that accepts an input of four values to produce an output of three values. It might appear to be a jumble of lines, squares, and circles, but the spirograph-like web connecting the circles and squares reveals structure. Indeed, we might notice that Figure 1.2 consists of layers, moving left to right, and that the layers are collections of neurons (nodes) like the one in Figure 1.1.

MLPs are fully connected, feedforward neural networks. Fully connected because every input to a layer is passed to every node of the layer, and feedforward because the output of every layer moves "forward" (to the right) to become the input to the next layer without a feedback loop where an output of a layer becomes an input to an earlier layer. The CNN architecture we're building toward is the same: information flows one way, from left to right, layer by layer, from the model's input to its output. Let's walk through Figure 1.2 beginning on the left.

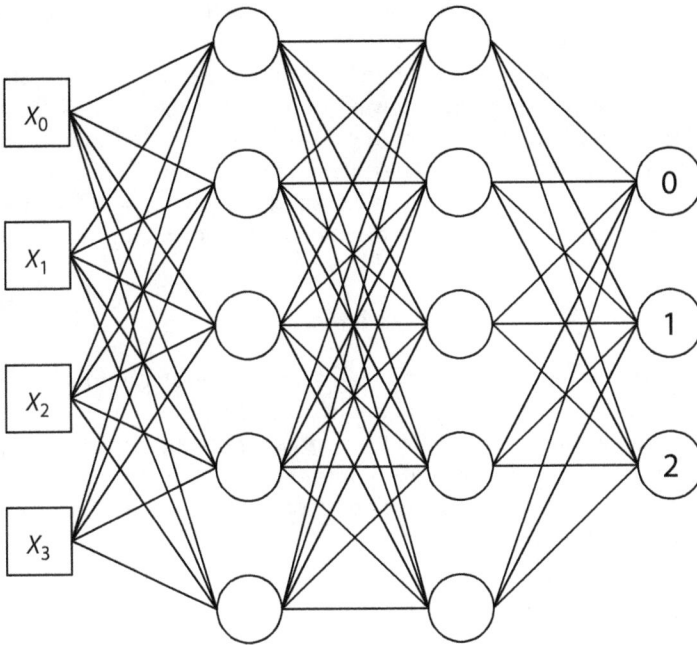

Figure 1.2 A multilayer perceptron (MLP)

The squares, labeled x_0, x_1, and so on, mirror the squares in Figure 1.1. Collectively, the squares form a *vector* known as the *feature vector*. The feature vector is the input to the neural network:

$$x = (x_0, x_1, x_2, \ldots, x_{n-1})$$

for an n-element feature vector, x. Notice that I'm indexing from zero, not one, as mathematicians do.

A quick comment on notation is in order. I'll use x for vectors and x for scalars. A scalar is a single number like 42, $\sqrt{2}$, or 8675309. Vectors are one-dimensional collections of scalars akin to one-dimensional arrays in programming, hence indexing from zero. Vectors have rich mathematics associated with them, but we'll ignore that richness and treat vectors as collections of numbers used by models. Later in this chapter, I'll talk about matrices (singular "matrix") which I'll represent with a capital letter like X. Matrices are essentially two-dimensional arrays of numbers. Standard practice represents a dataset as a matrix where each row is a feature vector.

The numbers in a feature vector are the *features*. Features represent characteristics of the thing-to-be-classified deemed useful for such a classification. In other words, given a set of features, we want the MLP to produce an output we can use to assign the thing described by the features to a particular class, a particular category like "sparrow" or "crow." Much of our effort in this book focuses on producing such feature vectors in two-dimensions, otherwise known as images.

As stated, information flows through the network from left to right. The input features (the xs) are multiplied by the weights associated with every line between the squares and the first layer of nodes (circles), exactly as presented in Figure 1.1. The products are then summed appropriately for each node, including each node's specific bias value (not indicated), to act as input to the node's activation function. Typically, each node uses the same activation function per layer and often for the entire network. The resulting outputs from the activation function become a new set of inputs for the second layer nodes, then the third, and so on until finally reaching the output layer. The layers between the input and output are known as *hidden layers*. The phrase "deep learning" refers to networks with many hidden layers, from dozens to over a hundred or more for some networks.

Mathematically, neural networks are functions accepting an input (x) and producing an output (y) with parameters (θ):

$$y = f(x; \theta)$$

for $f()$ a function that accepts a vector input and returns a vector output.

Training's goal is to set θ, a compact mathematical representation of all the weights and biases of the model. In the end, neural networks are nothing more than mathematical functions mapping inputs of some kind (represented by x) to outputs of some kind (y). For us, the inputs will be images of birds or birdsong, and the output will be a class label indicating the type of bird.

1.3.2   Convolutional Neural Networks

MLPs can be effective, but drawbacks limit their success, especially when working with inputs showing 2D structure. The primary limitation of MLPs is that they are parameter-hungry. A fully connected layer with n inputs mapped to the m nodes of the layer requires learning nm weights plus an additional m bias values.

The simple network of Figure 1.2 requires learning 73 parameters in total: $4 \times 5 = 20$ weights to map the input to the first hidden layer, plus 5 bias values, followed by $5 \times 5 = 25$ weights between the hidden layers and another 5 bias values to $5 \times 3 = 15$ weights from the second hidden layer to the output layer, plus 3 bias values:

$$4 \times 5 + 5 + 5 \times 5 + 5 + 5 \times 3 + 3 = 73$$

The network of Figure 1.2 is tiny, similar to the models people worked with in the 1970s and 1980s. As we'll learn later in the chapter, it's sufficient for a simple experiment but woefully inadequate for the kind of data we intend to exploit.

The fact that the number of weights between layers in an MLP grows as n^2 (assuming n and m are approximately equal) means that the complexity of an MLP in terms of the number of parameters to learn during the training explodes as the size of the layers increases. And, to make matters worse, a large initial input vector implies large numbers of hidden layer nodes to learn the desired mapping from input to output. Images are challenging because unraveling row next to row results in a massive input feature vector (the pixels of the image are the features).

For example, a small input image might be 128 rows by 128 columns, hardly more than a thumbnail. That's already an input vector of $128 \times 128 = 16,384$ features. Now, make that image a color RGB image, and we're up to an input feature vector of $128 \times 128 \times 3 = 49,152$ features. Such a large input requires a substantial number of hidden layer nodes to allow the model to learn a meaningful approximation of the (assumed) function mapping input images to output labels.

To make matters worse, the training data required for such a model explodes as the input feature vector length increases. Neural network training seeks to create a useful model using a collection of labeled samples as a proxy for all the data that could be generated by the parent distribution, the what-ever-it-is that creates images, in this case. As the number of parameters to learn explodes, the amount of training data to successfully capture that function explodes. This effect is known as the *curse of dimensionality*, and it stymied neural networks for decades.

I'm not done complaining yet. Images unraveled as feature vectors are bad enough for the reasons given, but an additional complication seems to doom our enterprise before it begins. MLPs work best when the features in the feature vector are *uncorrelated*. This means that as feature x_0 changes, features x_1 and, say, x_{13} don't also change in an easily predictable way. If they do, they are not informative to the model and inhibit learning. Indeed, early machine learning practitioners spent considerable effort seeking out correlated features and removing them from the feature vectors.

How does this relate to images? The pixels of the image are our features. Find an image and pick a random point (pixel) in the image. Now, look at the neighboring pixels. In most cases, on average, the neighboring pixels will be essentially the same color or intensity if grayscale. In other words, pixels in images are highly correlated. This is not good, and not hopeful for success with MLPs if we want to learn to identify what's in an image.

We appear to be out of luck. And so machine learning was for ages, until the late 1990s, when researchers – in particular Yann LeCun and his research group – got clever. They realized that it is possible to replace a sequence of fully connected MLP layers by convolutional layers that employ a classic digital image processing trick, *convolution*, thereby swapping the fully connected weight matrix with a smaller set of convolutional filters. The savings in weights was dramatic and had other advantages as well. This insight led directly to the introduction of CNNs and opened the door to the flood of AI that the world currently enjoys (and is slightly afraid of). MLPs are holistic: they interpret their input vector as a whole. As such, the structure within the input vector is "perceived" by the network at a glance, we might say. CNNs, on the other hand, pay attention to the structure in the input via the convolution operation. Because of this, a CNN can detect critical aspects of the input image that are essential to classification regardless of where they appear in the image. This talent proved to be the key to deep learning.

Let's proceed by first understanding what convolution is as it applies to CNNs and follow by understanding the structure of a typical CNN of the kind we will use throughout the projects.

Understanding Convolution

Convolution has a mathematical definition involving integrals. Fortunately for us, in practice, 2D convolution with digital images becomes a simple algorithm that slides a small kernel (think of a tiny 3×3 or 5×5 pixel image) over a larger image. Every pixel covered by the kernel has its value multiplied by the corresponding kernel value, and the entire set of products is summed and used as the value placed in the output image to match the location of the current kernel's center. Words make the process seem more complicated than it is. Figure 1.3 will help.

The figure contains three arrays of numbers. The array on the left is 8×8 and represents an input to the convolution operation (that is, an image). The 3×3 convolution kernel is in the middle, with the output of the convolution the 7×7 array on the right. We'll get to the empty boxes momentarily. Note that I've represented the convolution operation as *, which is sometimes used in the literature, though there is no hard and fast symbol.

3	1	4	1	5	9	2	7
2	7	1	8	2	8	1	8
8	6	7	5	3	0	9	0
3	9	9	1	6	8	0	0
8	5	3	9	7	3	4	2
1	4	1	4	2	1	3	6
6	4	8	0	7	4	0	7
0	3	1	8	3	0	9	1

$*$

-1	0	1
-1	2	1
-1	0	1

$=$

	13	2	14	7	18	0	
	16	6	4	8	-1	10	
	18	13	-1	8	13	-9	
	11	2	20	12	-2	4	
	5	2	12	-1	-7	13	
	11	17	2	7	8	9	

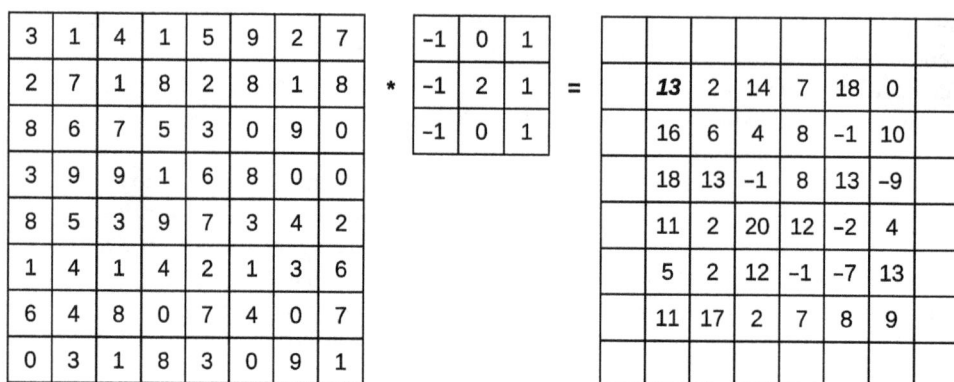

Figure 1.3 Convolution with a 3×3 kernel

To perform the convolution, imagine placing the kernel on top of the upper left corner of the input so that we have two 3×3 arrays, which we'll multiply together, corresponding element by corresponding element:

$$\begin{bmatrix} 3 & 1 & 4 \\ 2 & 7 & 1 \\ 8 & 6 & 7 \end{bmatrix} \odot \begin{bmatrix} -1 & 0 & 1 \\ -1 & 2 & 1 \\ -1 & 0 & 1 \end{bmatrix} = \begin{bmatrix} -3 & 0 & 4 \\ -2 & 14 & 1 \\ -8 & 0 & 7 \end{bmatrix} \rightarrow 13$$

The final 13 is the sum of all the elements in the 3×3 product. Notice that 13 is also the upper left value on the right in Figure 1.3. The product of the kernel elements and the portion of the input array it currently overlaps is used to replace the center of the kernel position in the output array. The $\odot$ symbol is typically used for elementwise multiplication, also known as the Hadamard product. The NumPy toolkit multiplies arrays this way by default.

To complete the convolution, slide the kernel one position (pixel) to the right and repeat updating the following position in the output array. At the end of a row, return the kernel to the leftmost position, slide down one position, then continue. The convolution is complete when the kernel has moved over the entire input.

The center of the 3×3 kernel position in the output is updated, implying that 2D convolution with a 3×3 kernel produces an output array with a one-pixel border where there are no values. Figure 1.3 illustrates the process usually known as *exact* or *valid* convolution. Edge pixels are typically assigned by imagining a border of constant value (usually 0) around the input. In that case, the edges of the output will be given a value, and the convolution operation on an $n \times n$ input will lead to an $n \times n$ output. In code, *zero-padding* the border in this way is often referred to as *same*, e.g., mode='same'. Most of the project CNNs use zero-padding.

Convolution is straightforward enough to understand, but why do it? The field of digital image processing learned early on that convolution is an excellent way to detect the presence of specific structures in an image. For example, consider Figure 1.4.

The figure illustrates the effect of different 3×3 convolutional kernels on the Northern Flicker image in the upper left. The top row emphasizes horizontal edges, followed by vertical edges. The bottom row emphasizes edges running from the lower left to the

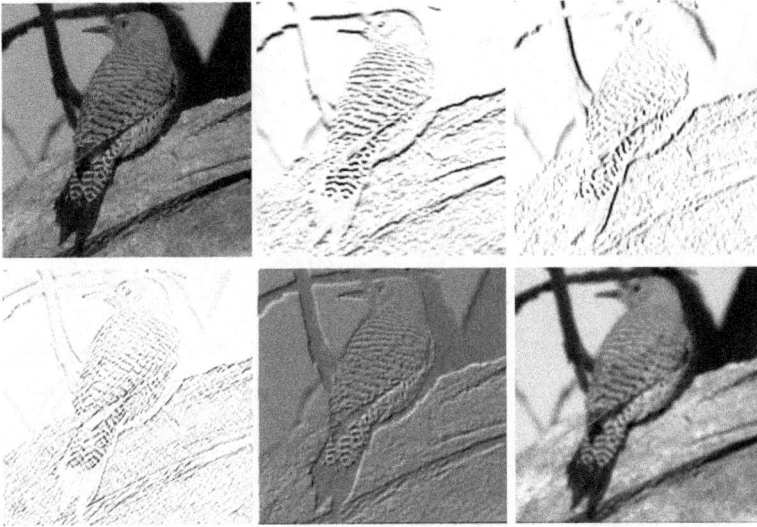

Figure 1.4 Convolution examples. Top: original, horizontal, vertical. Bottom: 45 degrees, embossed, smoothed

upper right, followed by embossing and smoothing to blur the image. Be aware that the horizontal, vertical, and 45-degree examples are inverted to be more visible in print.

The critical observation is that convolution produces outputs emphasizing specific aspects of the image. This is precisely what we want a CNN to do – transform the raw input image using kernels learned during training to detect image content helpful for classification. It isn't difficult to imagine a CNN learning kernels to detect the presence of the Northern Flicker's tail chevrons as indicative of the species. Of course, *how* it might learn such a thing is an entirely different matter, to which we now turn.

Understanding CNNs

Figure 1.5 conceptualizes a convolutional neural network similar to those we'll use later in the book. We must understand each part of the figure. Fortunately, none of it is particularly complex, at least conceptually. Implementation is another matter, and one so beset by "the devil is in the details" that only the boldest attempt to create robust CNN code from scratch. Most use one of the existing deep learning toolkits. In this book, we'll use Keras on top of TensorFlow. CNNs use the same parts in the same way, regardless of implementation.

Figure 1.5 should be read left to right, top to bottom. The rightmost part of the top row is duplicated as the leftmost part of the bottom row for convenience. A cursory glance tells us that the input to the CNN is an image, here a 128×128 pixel RGB image of a Wood Duck. The "128×128×3" text below the input image should be read as 128 rows (the height in pixels), 128 columns (the width), and 3 channels (red, green, blue). CNNs operate on *tensors*, which means multidimensional arrays of numbers. For example, most CNNs express tensors as (N, H, W, C) for N samples (images) of H height, W width, and C channels. The text below the input in Figure 1.5 is a single image ($N = 1$), implying no need to include N. The text at the bottom of each part of the figure similarly tells us the output of each CNN layer as (H, W, C). The CNN's layers accept the previous layer's

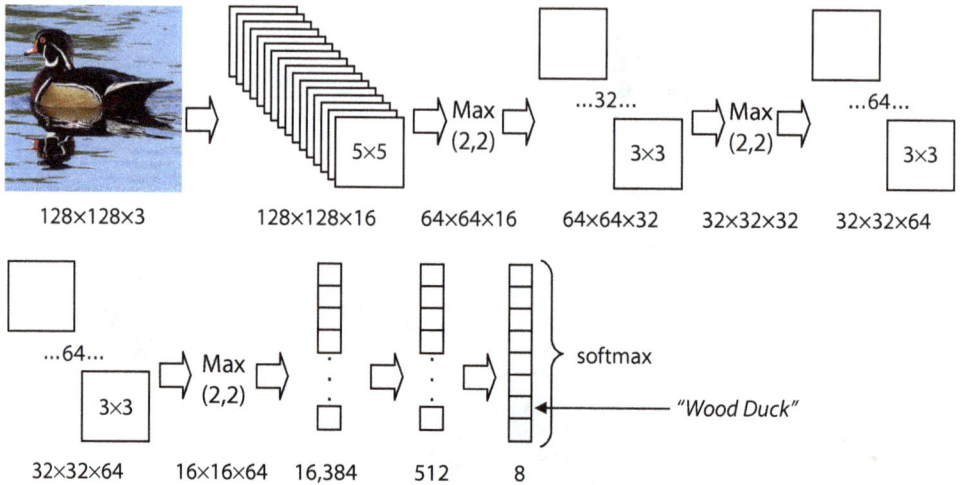

Figure 1.5 A conceptual diagram of a CNN

output tensor as input to produce a new output tensor, just as the layers in an MLP accept the previous layer's output vector as input to produce a new output vector.

To the right of the input we see a stack of squares labeled "5×5" with "128×128×16" below. If you care to count, you'll notice 16 squares in the stack. The stack represents a *convolutional layer*, one that learns 16 *filters* using 5×5 kernels to produce 16 output *activation maps*. Convolutional layers learn filters which are themselves 3D stacks of kernels. Each filter is used at inference time to produce an activation map. The set of activation maps produced by the filters becomes the input tensor to the next layer in the CNN. In Figure 1.5, the input to the first convolutional layer has 3 channels; therefore, each filter consists of a stack of 3 kernels. The kernels themselves are 5×5 pixels, as indicated. To apply the filter, convolve each channel of the input tensor with the corresponding kernel, then sum output pixels across the channels to produce a single output activation map. For the first convolutional layer in the figure, then, the first filter convolves each of the 3 input channels with the corresponding 5×5 filter, using padding to produce three 128×128 pixel outputs (one for each channel), which are then summed across channels to produce a single 128×128 output activation map. Repeating for all 16 filters produces the "128×128×16" output tensor indicated. The first convolutional layer here consists of 16 filters, each requiring a stack of three 5×5 kernels. Additionally, each filter typically includes a bias term adding 16 more parameters to the total. Therefore, during training, the first convolutional layer learns:

$$16 \times (5 \times 5 \times 3) + 16 = 1216$$

parameters. Notice that the calculation excludes the size of the input image. Convolutional layers do not pay attention to the spatial extent of the input; they only pay attention to the number of channels and the size of the kernels. The layer works just as well with a 1024×1024 pixel input to produce a 1024×1024×16 output tensor. This flexibility is not present in traditional MLPs.

What if we decide to replace the first convolutional layer with a standard feedforward layer accepting a same-sized input vector and producing a same-sized output vector? Doing so requires first unraveling the input image to form a $128 \times 128 \times 3 = 49,152$ element input vector mapped to a $128 \times 128 \times 16 = 262,144$ element output vector. The weight matrix between such layers implies learning $262,144 \times 49,152 = 12,884,901,888$ weights plus another $262,144$ bias values. This isn't feasible in practice, and alone is justification for using convolutional layers, to say nothing of the benefits of learning local feature detectors, that is, filters.

The subsequent layer in Figure 1.5 is marked "Max(2,2)." This is shorthand for a *pooling layer* which has no learnable parameters. Pooling layers reduce the spatial extent of a tensor by applying an operation over each channel of the input tensor. The "Max" designation means "maximum," and the "(2,2)" portion implies a 2×2 pooling.

Pooling moves over the input tensor's height and width, (H,W), separately for each channel. Each 2×2 region is replaced in the output tensor by the maximum value in that region. Unlike convolution, pooling does not overlap, in general, but moves block by block. For example, if the first two rows of a particular channel of the input tensor begin with (space added to separate the blocks):

$$8 \quad 10 \quad 3 \quad 0 \quad 7 \quad 7 \quad \ldots$$
$$3 \quad 8 \quad 9 \quad 4 \quad 5 \quad 3$$

then, a 2×2 max pooling operation looks at each non-overlapping 2×2 block and selects the maximum in that block to produce a single output row beginning with:

$$10 \quad 9 \quad 7 \quad \ldots$$

representing the largest value in each 2×2 block.

Therefore, a 2×2 max pooling layer reduces the spatial extent of each channel of the input tensor by a factor of 2 in each direction without changing the number of channels. This is reflected in the figure by mapping the 128×128×16 output of the first convolutional layer to a 64×64×16 tensor.

The CNN in Figure 1.5 contains two more convolutional layers using 3×3 kernels, each followed by a max pooling layer. These, collectively, transform the input like so:

$$64 \times 64 \times 32 \rightarrow 32 \times 32 \times 32 \rightarrow 32 \times 32 \times 64 \rightarrow 16 \times 16 \times 64$$

Notice that the number of filters in the convolutional layers increases as we move through the model. Experience teaches that such an approach is often helpful as it allows the model to learn a richer set of filters (kernels) as data flows through the model, thereby increasing the probability that meaningful filters will evolve during training. Two additional points are worth mentioning here. First, as we move deeper into the network (or, as illustrated, successively to the right), the portion of the original input image influencing each value in the tensor input to a layer increases spatially. This is referred to as the *effective receptive field*. Therefore, the features highlighted by filters in deeper convolutional layers respond to more spatially complex portions of the original input. The first convolutional layer typically learns filters corresponding to edges, textures, and colors, but by the second or third, the filters are learning about aggregates that might correspond, for a face input, to the eyes, nose, mouth, and, later, the relationship between them. These features enable the topmost part of the network (described below) to classify the entire image accurately.

The second point concerns *what* is learned by each convolutional layer. The filters learned by the first convolutional layer are little different from the sort of filters anyone experienced with digital image process might use. What, however, is the meaning of the collection of filters learned by, say, the third convolutional layer? The operation performed by each layer is identical, but the meaning of the filters at deeper layers is unclear and probably uninterpretable by humans. This, to me, is a fascinating state of affairs and a pale reflection of what happens with large, generative AI models like ChatGPT, which are far beyond human comprehension in terms of understanding the meaning within the data passed through the model.

Let's return to the figure. After the last max pooling layer comes a *flatten* layer, which unravels a 3D input tensor to produce a vector. The input tensor is $16 \times 16 \times 64 = 16,384$ elements, implying an output vector of $16,384$ elements. Given the comments of the previous paragraph, we're now in a position to appreciate what the flattened output of the convolutional and pooling layers is: a new and abstract representation of the original input image, one that contains, in a compressed numerical form, the essential characteristics of the input image.

The flattened output is passed to a *dense* layer with 512 nodes. Dense layers are nothing more than traditional fully connected layers in disguise. Therefore, the weight matrix between the output of all the convolutional and pooling layers and the dense layer contains:

$$16,384 \times 512 + 512 = 8,389,120$$

weights and biases. For many CNNs, especially those we'll explore in this book, most of the model's weights are in the dense layers at the top.

The final layer is labeled "softmax" and has 8 nodes. This is the model's output. The fact that there are 8 nodes means the model is trained to recognize 8 different input classes, one corresponding to "Wood Duck." Multiclass models typically use softmax outputs. The softmax is a generalization of the logistic operation used by binary models that separate inputs into "this" or "that" class. For us, the softmax can be viewed as a vector of probabilities (likelihoods), each representing the model's belief that the input is a member of that class. We usually select the most significant softmax output and return the corresponding class label. If the model illustrated in Figure 1.5 is well trained, it will produce an 8-element softmax vector for the Wood Duck input such that the vector element matching the Wood Duck label is the largest. The sum of the elements of a softmax vector is always 1.0, which explains the probabilistic interpretation. However, some care is required because most deep learning models are overly optimistic, and the softmax output isn't a true probability of class membership but is best understood as a belief or likelihood. In practice, however, people are often somewhat sloppy and speak of probabilities, knowing that the listener understands the nuance involved.

We have arrived at a basic understanding of the types of models we'll use in the projects. Let's now turn our attention to data. Ultimately, the entire AI enterprise stands or falls on the quality of its datasets.

1.4   Datasets, Training and Testing

It's time to get practical. The core of our effort is focused on constructing datasets we use with existing toolkits to build working neural network models. This section walks through an example to familiarize us with the essence of the process.

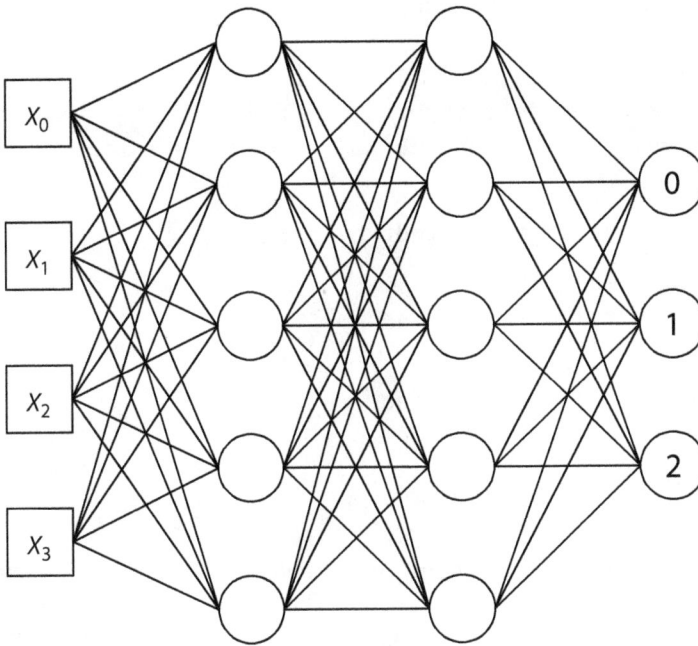

Figure 1.6 A multilayer perceptron (MLP)

Let's create the network of Figure 1.2, reproduced here as Figure 1.6. We'll use Python, the programming language at present most often associated with AI. Then we'll train and test network using known inputs not seen during training.

The example uses Scikit-Learn, a standard machine learning tookit. Scikit-Learn isn't the toolkit we'll use for our experiments, but it's helpful for this example because it creates MLPs with minimal effort. The curious will find the source code in the file *iris_mlp.py*. To install the toolkit, open a command prompt and enter:

```
> pip3 install scikit-learn
```

Chapter 3 builds the environment, but the command above should get you started if you're interested in exploring the iris example in detail. Our network expects four-element feature vectors. By pure coincidence, a famous machine learning dataset built into Scikit-Learn consists of four measurements of three species of iris flower. The dataset consists of 150 measurements of individual flower petal and sepal length and width in centimeters. The source code reserves 100 for training the network and 50 for testing. The flower classes are identified by integers, 0, 1 or 2, which is a standard approach. We don't care about the particular species in this example. However, notice that there are three classes of irises and three outputs for the network. Each output represents the model's belief, [0, 1], that the input represents a member of that class.

Let's walk through the code. The fact that the architecture isn't a CNN does not alter the utility of the example; the pieces remain the same for the more complex networks we'll use in the projects.

First, let's load the data and split it into train and test datasets:

```
import numpy as np
from sklearn.datasets import load_iris
from sklearn.neural_network import MLPClassifier

iris = load_iris()
x,y = iris.data, iris.target

np.random.seed(8675311)
i = np.argsort(np.random.random(len(y)))
x,y = x[i],y[i]

xtrn, ytrn = x[:100], y[:100]
xtst, ytst = x[100:], y[100:]
```

Listing 1.1 Building the iris dataset

Listing 1.1 is split into four sections, "paragraphs" as I'll refer to them. Here's where I'm assuming some familiarity with Python. The first three lines import the NumPy library, universally referred to in code as np, along with portions of the Scikit-Learn library (sklearn in code).

The load_iris function loads the iris dataset included with Scikit-Learn. You may see a message about downloading the data on your first run. The second code paragraph loads the dataset, then extracts the data itself (iris.data) along with the class labels (iris.target) as x and y. What do x and y look like?

Stopping the code and examining x and y reveals the following:

```
>>> x.shape
(150, 4)
>>> x[:5]
array([[5.1, 3.5, 1.4, 0.2],
       [4.9, 3. , 1.4, 0.2],
       [4.7, 3.2, 1.3, 0.2],
       [4.6, 3.1, 1.5, 0.2],
       [5. , 3.6, 1.4, 0.2]])
>>> y.shape
(150,)
>>> y[:5]
array([0, 0, 0, 0, 0])
```

Here, >>> is Python's interactive prompt indicating places where I entered code with Python's response following immediately. Both x and y are NumPy arrays, the first a matrix of 150 rows of 4 columns and the second a vector of 150 elements. This is what the shape attribute tells us.[1]

---

1 NumPy is essential to everything we do. Look for *tutorial.pdf* on the book's GitHub site for a brief tutorial.

To examine the first five feature vectors, use x[:5]. The result is a matrix of 5 rows, each with 4 columns. The first feature vector, [5.1, 3.5, 1.4, 0.2], tells us the measurements of the first flower's sepal length and width followed by its petal length and width. We aim to train the network to accept this vector and produce an output that leads us to select the correct corresponding class label. The actual class labels are in y, the first five of which are class 0, meaning the first five feature vectors refer to flowers of the same class. As it happens, the iris dataset is returned in class order so that the first 50 are from class 0, the next 50 from class 1, and the final 50 are from class 2. And here's our first example of the necessity of data preprocessing. We intend to train a model to accept unknown flower measurements and, hopefully, return a proper class label. To do that, we use existing data as the *training set*. All our data about iris flowers is presently in x with matching labels in y, so we must construct our training set from the two variables. Further, to test whether our model is performing well, we must have a *test set*, a collection of samples for which we know the correct answer but were not used to train the model. We must also select our test set from x and y.

Let's assume we want to use $\frac{2}{3}$ of the data for training and hold back the remaining $\frac{1}{3}$ for testing. If we are quick to code, we might keep the first 100 feature vectors and labels for training and the final 50 for testing. However, Scikit-Learn has returned the dataset to us in an ordered fashion, so following this plan will lead to disaster: we'll train the network on examples of class 0 and class 1 and then attempt to test it with examples from class 2. Failure is assured – the model never saw instances of class 2 during training, so how could we expect it to identify members of that class?

This observation is more than academic; it's critical to our entire research enterprise. Neural networks are trained on data under the assumption that the data fairly represents what the model will encounter when it is used in the wild. Therefore, training the model with a dataset that's representative of the full range of data it will encounter means the difference between success and failure.

There are several ways to think about what is happening if the data isn't a fair representation. One way relies on the difference between interpolation and extrapolation. We want models that are, in a vague sense of the word, interpolating when we use them. Such models work with a range of inputs similar to those encountered during training, so we might reasonably expect them to succeed at inference time with similar inputs.

Extrapolation means going beyond the existing data. The neural networks we'll train in this book are poor extrapolators. Practitioners refer to inputs beyond the sort the model was trained to digest as *out of distribution*; inputs fundamentally outside the range used to condition the model. Extrapolating is to be shunned.

The previous paragraph used the word "distribution." In this sense, it refers to a probability distribution, where the distribution is the possible set of data (feature vectors) that the model might encounter. I think of it Platonically: there is an ideal parent distribution, a data generating process that produces feature vectors, for example, iris flower feature vectors. When we train the model, we use a sample of such feature vectors with labels to represent the whole. The better our sample is at representing all that the parent distribution can generate, the more successful the model will be when we use it in the wild.

Let's return to the task at hand. We have our full dataset in x and y, but we know it's ordered. We don't want that. We want to split the dataset between train and test so that each is (statistically) similar, especially by class label. There are pedantic ways to do this, ways than ensure different kinds of balance, but much of the time all we need do is stir the pot by randomly shuffling the order of the feature vectors with labels following their proper feature vector. This is the job of the third code paragraph in Listing 1.1:

```
np.random.seed(8675311)
i = np.argsort(np.random.random(len(y)))
x,y = x[i],y[i]
```

We'll encounter code like this throughout the book, so I'll describe it in detail. The first line is for your convenience. It sets the NumPy pseudorandom number seed in an old-fashioned way you wouldn't use in production code. Setting the seed ensures that the "randomness" used to shuffle the dataset is repeatable.

The second line creates a random vector containing the integers 0 through 149, precisely the range of indices we might use to index the 150 samples stored in x and y. It does this by combining two NumPy functions. The first returns a 150-element (`len(y)`) vector of random floating-point numbers in [0, 1]. This vector is then passed to np.argsort which returns the set of indices that would sort the vector instead of the sorted vector values themselves. This set of indices is a scrambling of 0, 1, 2, …, 149 where each number appears only once.

The final line is a grouping of two statements: x=x[i] and y=y[i]. In NumPy, indexing with an array, here i, returns a new array in the order the indices appear. This returns a new ordering of x and y, thereby mixing the previously ordered data set.

Completing construction of the train and test datasets is now merely a matter of selecting the first 100 or the final 50,

```
xtrn, ytrn = x[:100], y[:100]
xtst, ytst = x[100:], y[100:]
```

giving us a training set (xtrn, ytrn) and a test set (xtst, ytst). For example, xtrn now begins like so:

```
>>> xtrn[:5]
array([[5.7, 4.4, 1.5, 0.4],
       [5.1, 3.3, 1.7, 0.5],
       [5.7, 2.6, 3.5, 1. ],
       [5.5, 3.5, 1.3, 0.2],
       [7.3, 2.9, 6.3, 1.8]])
>>> ytrn[:5]
array([0, 0, 1, 0, 2])
```

telling us that of the first 5 training samples, three are from class 0 with one each from class 1 and class 2. Fixing the pseudorandom number seed ensures you will see the same if you run *mlp_iris.py*.

I've been somewhat pedantic with this example, but the extra attention is worthwhile in the long run. Building datasets and understanding what goes into a good dataset is critical to deep learning success. Everything is data. The old adage "garbage in, garbage out" has never been more true.

We have train and test datasets. All that remains is to specify the model (recall, we're training the architecture in Figure 1.2), train it, and then test it. Here's where the power of toolkits comes in handy. For Scikit-Learn, defining the model, and then training takes all of two lines of code:

```
mlp = MLPClassifier(hidden_layer_sizes=(5,5), max_iter=2000)
mlp.fit(xtrn,ytrn)
```

The first line creates an instance of `MLPClassifier`. This is the model. To specify the architecture, we only need to indicate the number of nodes we want in each hidden layer.

Figure 1.2 has two hidden layers, each with five nodes, hence `(5,5)` as the argument to `hidden_layer_sizes`. The extra `max_iter` keyword tells the model to train for up to 2000 epochs. An *epoch* is a complete pass through the training set in the sense of showing the model each of the training samples and determining how to update the model's weights and biases based on the accuracy of its output compared to the known training set labels.

The call to `fit` trains the model. We pass in the training set (the collection of feature vectors) and the corresponding class labels, Scikit-Learn takes care of the rest, including understanding that feature vectors have four elements (the four inputs to the model), and there are three classes, implying three outputs from the model.

The connections between the nodes of Figure 1.2 are the network weights. We need these weights, and the biases for each node, to train the model for the task at hand. These are the magic numbers that make the network perform properly. Finding them involves a series of steps that collectively fall under the umbrella of *optimization* – we start with a guess, then do things to make the guess better and better until we reach a point where we declare the model "trained." The call to `fit` hides much. On the one hand, that's good for us in practice. On the other hand, it's not quite so good because we might be tempted to gloss over important elements of the training process. What `fit` is hiding has to do with two key algorithms, algorithms that apply to virtually all neural networks: *gradient descent* and *backpropagation*.

Gradient descent can be thought of as going downhill from a higher point to a lower point, ideally, the lowest point possible. Optimization problems do this to find the best set of values for the problem. In the world of neural networks, the best set of values refers to the best set of weights and biases that lead to the smallest number of mistakes when classifying the training samples.

The network is initialized, intelligently, with small random values. This corresponds to starting from some point above the lowest point. Gradient descent then steps in the direction leading most steeply downward from the current point. Mathematically, this is in the direction opposite to the maximum gradient, with *gradient* a multidimensional analog of the slope of a line on a curve at a specific point. Moving down in this way, calculating a new gradient direction after each step, will tend to move from a higher point to a lower point – from greater error on the training samples to less error.

Over time, gradient descent will reach a minimum. It might be the best possible minimum leading to the smallest training set error (the *global minimium*), or it might be a *local minimum*, a point from which gradient descent cannot climb out of. For many neural networks, it is believed that there are many local minima and that, in general, most of them are about as effective in practice as any other. That's the hope, at any rate.

Gradient descent trains the neural network by updating the weights and biases on each step. A natural question is: how much should the weights and biases be changed? The answer comes from calculus, from the process of computing the gradient. What is needed is a sense of how much the error, known as the *loss*, is affected by a slight change in each weight and bias of the network. In calculus terms, we must learn the partial derivative of the full loss function with respect to the weights and biases – enter backpropagation. Backpropagation is a clever and sometimes complex application of the chain rule for derivatives. With backpropagation, we learn how a slight change in a weight affects the loss for the current state of the network's weights and biases and the current estimate of the model's error on the training set. Gradient descent updates the weights and biases by a fraction of this value, step by step, to move the network closer and closer to a state where it acts as we wish. When the model performs satisfactorily, we declare it trained and say it's ready for use in the wild to classify new, unknown inputs.

Let's bring things back to our example. When the call to `fit` returns, all that the previous paragraphs imply has occurred, and Scikit-Learn considers the network to be trained. I am ignoring additional aspects of the learning process, in particular important settings used to guide the training process, *hyperparameters*, but for now, that's just fine. We will return to them when as we work through the projects. The `fit` method has trained the MLP. Let's use it on the held-out test set to measure how well the model has learned:

```
prob = mlp.predict_proba(xtst)
pred = np.argmax(prob, axis=1)
```

Two variables are created: `prob` and `pred`. The first comes from the `predict_proba` method which accepts the test inputs (`xtst`). The output is a NumPy matrix with as many rows as there are rows in `xtst` and as many columns as there are classes, here three. For example, the first five rows of `prob` are,

```
0.0010 0.9717 0.0273
0.0015 0.9937 0.0048
0.0006 0.0909 0.9084
0.0007 0.9941 0.0053
0.0008 0.9744 0.0249
```

Each row here is the model's assigned likelihood of each class membership for the corresponding test sample in `xtst`. In this case, the model indicated a likelihood of 0.0010 that the first test sample is a member of class 0, a likelihood of 0.9717 that it is a member of class 1 and a likelihood of 0.0273 for class 2. The sum across the rows is always 1.0. Therefore, the model is quite sure that the first test sample is an instance of class 1 since that's the greatest likelihood value. The model is correct in this case. It believes the same about the second test sample but assigns the third to class 2.

The `pred` array holds the label associated with the largest value across the columns of `prob`. In other words, it holds the model's predicted class. Is the model correct most of the time? The remainder of the code in *iris_mlp.py* displays, for each test sample, the model's likelihoods (formally, the *softmax* output) along with the predicted class label, the actual class label from `ytst`, and `True` or `False` based on whether the two are the same. If you examine the output, you'll notice that the model made one mistake by assigning an instance of class 1 to class 2.

There are 50 test samples, 49 of which were correctly classified for an accuracy of $49/50 = 0.98$ or 98 percent, as reported.

The output concludes with a curious 3 by 3 display:

```
[[13  0  0]
 [ 0 15  1]
 [ 0  0 21]]
```

This is a *confusion matrix*, and we will become quite familiar with them as we work through the projects. The rows of the matrix represent true class labels, while the columns are the class labels assigned by the model. For example, the first row corresponds to class 0, of which there were 13 instances. The model assigned all 13 instances to class 0, thereby explaining the 13 in the first column of the first row. Now consider the second row, which corresponds to class 1. There were 16 class 1 instances in the test set and 15 of them were assigned to class 1 by the model. The remaining instance was assigned to class 2, hence the "1" in the third column of the first row. Finally, all 21 class 2 instances were correctly assigned to class 2.

A confusion matrix visually represents the model's performance on test data. If the model is perfect, all test samples will be placed in the correct category and the matrix will be purely diagonal. Counts in off-diagonal elements represent errors, mismatches between the true class label and the label assigned by the model. It might help to think of a confusion matrix as a two-dimensional histogram, as it's generated in much the same way in code:

```
cm = np.zeros((3,3), dtype="uint8")
for i in range(len(ytst)):
    cm[ytst[i],pred[i]] += 1
print(cm)
```

The confusion matrix is cm, a 3×3 matrix because we have three classes. It is, by default, initialized to zero. The dtype keyword tells NumPy to treat the array as an 8-bit unsigned integer solely for display purposes. Unlike Python proper, which only distinguishes between integers and floating-point numbers, NumPy cares about classic data types.

The for loop indexes the actual labels in ytst and the assigned labels in pred. The body of the loop increments the elements of the matrix for whatever values the two arrays have at that paired index, taking care to make the row the true label. When the two match, a correct classification, the values are the same and one of the diagonal elements is incremented. Errors increment off-diagonal elements.

The confusion matrix is the first and often best way to assess the raw performance of a model. I'll introduce other metrics derived from the confusion matrix as we work through the projects.

Before continuing, we must discuss the role of randomness in building and training neural networks. The iris dataset used randomness to split the existing data between train and test sets. I fixed the pseudorandom number seed so that you get precisely the results indicated above. If you alter that seed or comment out the np.random.seed call, each run of *iris_mlp.py* will produce a separate split of the existing data, resulting in different performances on the test set.

For example, I commented out the line and executed the code multiple times resulting in accuracies ranging from a high of 100 percent (perfection on the test set) to a low of 70 percent with most results in the 96 to 100 percent range. The low accuracy was due (in part) to a poor mix of train and test data. The training data might have too few samples of a particular class for the model to learn about the class, or the test data might have many instances of a class the model has difficulty classifying correctly. We must be aware of this effect when building datasets. For the projects, properly constructed datasets, meaning sensible in their content, are critical to success.

There is another source of randomness at work in the model. Earlier, I stated that models are trained via gradient descent beginning with randomly assigned weights and biases. Machine learning researchers have spent considerable effort determining the best way to select these initial values, and Scikit-Learn implements the fruits of that research, but in the end, random is random. Gradient descent is often more successful when beginning at one location versus another. If the initial set of weights and biases is poor, the model might arrive at an inferior local minima and perform poorly.

Using np.random.seed before splitting the iris data into train and test sets forces not only the same split each time but also sets up Scikit-Learn to use the same initial set of weights and biases. To illustrate the effect of differing network initialization on each run while preserving the same data split, add np.random.seed() (with no argument) just before the call to MLPClassifier. This changes the model's initialization on each run by forcing NumPy to seed the pseudorandom generator from the computer's current state.

I added the line and ran the code multiple times. Sometimes, the output matched giving a model that was 98 percent accurate on the test set. Sometimes the model was 100 percent accurate, but sometimes it was only 58 percent accurate, and once only 38 percent. The datasets were the same each time; only the model's initial weights and biases changed.

To sum up, randomness appears in at least two places when building models: (1) when splitting existing data into train and test sets and (2) when initializing the model prior to training. We'll learn of a third place later when we build project models, but these are the two places we must never forget. Proper dataset construction mitigates the first effect, while repeated model training (if possible) can mitigate the second. However, randomness in the order in which training data is presented to the network often remains.

The iris example has introduced us to most of what we'll use in the projects regarding the process behind datasets, training and testing. Sometimes, we might introduce a third dataset, a *validation* dataset, but I'll describe such and why when necessary.

We're now ready to walk through the process we'll use to build the neural networks that we explore in the projects.

2. The Process

The complexity behind building successful deep learning models lies somewhere between quantum physics and baking a cake, leaning decisively towards the latter. This chapter aims to lay out "The Process," a set of guidelines (not rules) to help you design a successful model. Let's begin with the steps, then discuss them, noting additional factors to keep in mind. The assumption is that the end goal is already decided; we know what we want the model to do (inputs and desired outputs). The process then becomes:

1. Collect a large enough labeled dataset consisting of inputs and expected outputs.
2. Look at your dataset and decide how best to preprocess it.
3. Split the dataset into two or three disjoint subsets: train, test and (optionally) validation.
4. (Optional) Augment the subsets independently by transforming samples into plausible representations of the same class.
5. Select a model architecture and associated hyperparameters.
6. Train the selected model using the training set.
7. (Optional) Use the validation set, if any, to evaluate a trained model to iterate over architecture and hyperparameter adjustments.
8. Test the final model using the held-out test set.
9. Deploy the model for use in the wild.

The Process hides significant detail. Let's dive deeper.

2.1 Data Collection

Step 1: Collect a large enough labeled dataset

We intend to train a model using data as a proxy for the parent distribution, the thing that generates samples of the kind we want the model to interpret. Therefore, it makes sense to collect as large a dataset as possible, with suitable caveats. When I'm asked how much data I'll need for a particular task, my tongue-in-cheek reply is always the same: "all of it."

However, the dataset must be sensible. If we want a classifier that identifies American Crows, we could collect a bunch of crow images (class 1) along with images of Border Collies (class 0) and use that as our dataset. The resulting classifier will likely be sterling on the test set and correctly classify virtually all input crows, but we shouldn't expect it to be of much value when used in the wild. What if the input is a Common Raven, for example? The model will most likely label it as a crow because it never saw ravens in the training set as a counter-example.

Adding negative examples that appear similar to a particular class is a good idea to force the model to learn the differences between them. Such examples are often referred to as *hard negatives*. Adding Common Raven images as class 0 (not crow) would help the model learn more about "crow-ness" as something distinct from "raven-ness."

A good dataset has enough examples of each class, ideally in the same relative frequency in which they appear in the wild, to allow the model to learn the distinguishing features of the classes. This statement implies that the samples are suitably diverse regarding pose, orientation, lighting, and resolution. For example, if the model is used in a setting where inputs might be photos taken at 1 meter out to 10 meters away, then the training set must include such, for each class, to teach the model that sometimes the target is near and sometimes it is far, but regardless, it's still an example of that class. If all the training images were taken in the summer, don't expect the model to do well with winter scenes unless such are included in the training set.

The more constrained the possible inputs to the model, the better. In an industrial setting where we need to quickly examine photos of bottles on a production line is an easier task than placing a trail camera somewhere and dealing with whatever image it produces (the desired target, a spider in front of the lens, rain, snow, late night revelers, etc.). Let the expected diversity of inputs and relevance to the task guide dataset construction. I mentioned that the dataset should reflect the relative frequency with which each class appears in the wild. This frequency, known as the *prior probability*, is essential, but we must make trade-offs from time to time. If the target class is rare, we still need to have enough examples in the training set for the model to learn to identify it.

Suppose the target class appears on average once in every 1000 inputs to the model. In that case, a model that always replies "not the target" will be correct on average 999 times out of 1000 or 99.9 percent of the time. Such a model, while highly accurate, is useless. A training set with a 1000:1 "not target." to "target" mix will, at least initially, produce such a model. It's possible over training time that the model will eventually "get it," but the probability of that happening decreases with the number of target samples, to say nothing of the inherent randomness involved in model initialization leading to one training run that may learn while most fail.

The best way to manage imbalanced training data, meaning differing frequencies for the different target classes, is an old and still somewhat unsolved issue in machine learning. Fortunately, deep learning models seem less susceptible to imbalance, within reason, enabling us to "get away" with training until the model is good enough on every class. Still, we must be aware of imbalance and seek to counteract it by collecting more samples of rare classes, if possible. An ad hoc guideline I've used is a 10:1 ratio between common or negative classes and the target class if the target class is rare, even if the actual ratio is closer to 1000:1 or even 10,000:1. If collecting more data isn't plausible because it is rare, expensive or otherwise difficult to come by, then we must resort to techniques like data augmentation (Step 4). Models intended for what in our case might genuinely be considered "use in the wild" often include a "none of the above" class, alternatively known as a *background class* or a *NOTA class*. We want to assign birds to known categories, but we understand that some inputs to the model will not contain a bird at all. Adding a background class filled with diverse instances of non-bird images will help the model by giving it an out if the input isn't a bird. Otherwise, the model will assign likelihoods to all the possible classes it knows, and one of those will, by chance, be slightly larger than any of the others, leading to an erroneous class assignment. In other words, the model must choose, so giving it a NOTA class option leads to better performance with desired classes.

The dataset is critical, the very heart of the process, but it is also task-specific, so all I can do in this section is lay out some guidelines and mention this or that caveat. In the end, experience helps, as does remembering that the training set should be considered a proxy for all the model might encounter. The more complete the training set in that regard, the better the model's chance of success.

2.2 Data Preprocessing

Step 2: Look at your dataset and decide how best to preprocess it

Machine learning models differ in what they expect as inputs. We already know that traditional MLPs expect feature vectors and that convolutional neural networks are happiest (in this book) with images. However, there's more to it than merely passing a set of images to the model.

First and foremost — *look at your images*. I'm often surprised by the amount of time wasted that might have been recovered by simply looking at the images in the dataset. I'm even more dismayed when the time wasted is due to not following my maxim because I'm in a hurry (or forgetful). So, look at your dataset and convince yourself that it makes sense. Are there blank images or images so substantially different from what the model will encounter that they are not suitable? Are the images noisy or full of artifacts? If so, will the model encounter the same when used? If not, remove the uncharacteristic images. Common sense, informed by the end goal, guides the process.

2.2.1 Image Geometry

Practical input concerns include image size, aspect ratio, and color space configuration.

Many deep neural networks expect inputs to be of the same size, typically square, and usually rather small, say 224 by 224 pixels. If using such an architecture, the raw images must be resized and likely cropped. There are architectures that accept images of any size, but most of our experiments will employ same-sized input images.

Aspect ratio refers to the width and height of an image in some units, usually pixels. If the dataset's raw images have different aspect ratios, the usual approach is to crop by selecting a suitably sized section of the image.

For example, suppose we're building a dataset for a model that expects a 224×224 pixel RGB image as input and we want to use the raw Say's Phoebe image on the left in Figure 2.1. In this case, it is possible to capture the bird with a square crop, as shown on the right. The raw image is 3262×2334 ($W \times H$) pixels, and the crop is 1600×1600 pixels square. The model expects 224×224 pixel inputs, implying the cropped image is also resized to map from 1600 pixels to 224 pixels. This is a reduction of about 7×, so we should expect fine details to be lost. If fine detail is essential to distinguishing a Say's Phoebe from an Eastern Phoebe, we might be out of luck using such small model inputs.

However, cropping and resizing will sometimes not capture the entire subject. In those cases, I typically resize the image to align the raw image's smallest dimension with the corresponding model input size, then crop.

Figure 2.2 presents a 3102×2824 pixel raw image on the right. A simple crop will not capture the entirety of the Cooper's Hawk. However, if we first resize to map the smallest image dimension from 2824 pixels high to 224 pixels, the new image width is 246 pixels. We then select a center 224×224 crop to form the training set image as shown on the right

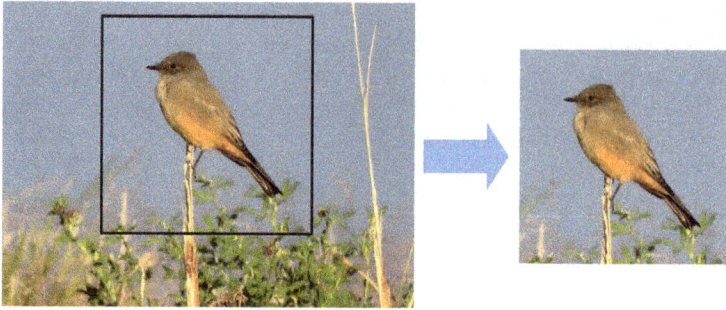

Figure 2.1 Cropping a raw image.

3102 x 2824   1758 x 1600

Figure 2.2 Rescaling before cropping.

(black square). The image in Figure 2.2 was resized to 1600 pixels, not 224, for illustration purposes.

We're beginning to appreciate the amount of effort that's sometimes required to build a good dataset, and we have more to contemplate. Fortunately, in practice, many of the images we want will be similar enough in size and aspect ratio that a consistent approach to selecting and cropping will capture almost everything without intense per-image processing. It won't matter too much if a few training images are poorly framed if we have enough of each class.

Color space configuration refers to the ordering of the color information in the input images. Digital color images of the kind we usually work with consist of red, green and blue information, [0, 255], for each pixel. These form the channels of the image where convention places the red channel first, followed by the green and blue, that is, as RGB. Some pretrained models expect the channels in a different order, perhaps BGR. If that's the case, the selected and cropped training images must then be shuffled to the proper color order. If not, the colors will be off, making the Bullock's Oriole of Figure 2.3 appear blue.

We're not entirely done with color spaces. In fact, there is much more we might say if we dive into the particulars of RGB versus standard RGB, to say nothing about other color formats. However, two issues are more likely to affect us: alpha channels and grayscale.

I've referred to images as RGB, but many images have an additional channel, the alpha channel: RGBA. The alpha channel isn't a color channel but a transparency channel used to denote the transparency of each pixel. Our models make no use of this information,

Figure 2.3 RGB channels (left) read as BGR confuses colors (right).

so it must be removed when building datasets. We'll use the Pillow library to read and write images from disk, which is fortunate, because Pillow makes it easy to convert input images between formats. For example, to read *oriole.png* and ensure that it does not contain an alpha channel use

```
image = Image.open("oriole.png").convert("RGB")
im = np.array(image)
```

where `Image` is a Pillow module and `np` references NumPy to transform the Pillow object into an array, in this case, one with three channels, not four: $H \times W \times 3$.

Grayscale images are single-channel images, with each pixel representing light intensity. Many models work with grayscale images, often requiring manual addition of the channel portion of the input. Transforming images to grayscale, regardless of their original configuration, is similarly straightforward using Pillow:

```
image = Image.open("oriole.png").convert("L")
im = np.array(image)
```

where `im` is now a two-dimensional array, $H \times W$. The `"L"` argument to `convert` stands for "luminance" or "luma." Pillow generates the luminance value from each pixel's RGB:

$$L = 0.299 \times R + 0.587 \times G + 0.114 \times B$$

The conversion formula reflects the sensitivity of the human eye to give more weight to green and less to blue.

2.2.2 Normalization and Standardization

Neural networks function best when their inputs are continuous and of similar magnitude without deviating significantly from zero. RGB and grayscale images are typically stored as bytes, [0, 255]. That every pixel falls into this range gives us the similar magnitude the model expects, but the magnitude itself is still too great; we want the input values to be closer to zero.

The most straightforward approach divides the input image by its largest value, usually its largest possible value of 255, which maps [0, 255] → [0, 1]. This process is called *normalization*, and we should do at least this much to the training data and at inference time. Whatever we do to preprocess the training data must be duplicated with any model input at inference time. For modern networks, normalization of pixel values is usually all that's required. However, it is entirely possible that we find ourselves working with advanced cameras that acquire more than 8 bits per pixel. Many CCD cameras produce 10-bit ([0, 1023]) up to 16-bit images ([0, 65535]). Normalization is still possible by dividing by $2^n - 1$ for an n-bit camera.

There are alternatives to simple normalization. One is *standardization*, which subtracts the mean of each feature before dividing by the standard deviation:

$$x \leftarrow \frac{x - \bar{x}}{\sigma}$$

Standardization is most effective for traditional feature vectors where the relative range of each feature might differ significantly. For images, standardization is usually unnecessary.

Early CNN toolkits encouraged users to create mean images found by literally summing every image in the training set before dividing by the number of training images. The mean image was then subtracted from each training sample to produce a new image representing the difference from the mean, a set of positive and negative values centered about zero. In practice, [0, 1] normalization is effective, but *mean image subtraction* is worth remembering for situations where only a portion of the input is likely to contain the subject with the remainder essentially static. I'm thinking here of a fixed camera observing a bird feeder or a particular view of a trail. Subtracting the mean image might encourage the network to focus attention on learning what has changed, input to input, though I suspect models are often able to discern that portions of the input are uninformative without mean image subtraction. Deep learning is an empirical discipline – when questions arise, the usual answer is "try it and see."

2.3 Data Splitting and Augmentation

Step 3: Split the dataset into two or three disjoint subsets
Step 4: (Optional) Augment the subsets independently

Constructing the training and test sets implies partitioning the dataset. How best to do it? How much should be used for training, and how much for testing? And, what's this mysterious "validation" set mentioned at the beginning of the chapter? Let's find some answers, then proceed to Step 4, data augmentation.

2.3.1 Train, Test and Validation

The training set is used to condition the model, meaning to determine the weights and biases of the network that lead to a model performing well enough for our purposes. The training set is a proxy for all the data that might possibly be encountered by the model, all the parent data. Therefore, putting most of the available data into the training set makes sense.

Classical machine learning often suggested an 80/20 split - 80 percent for the training set and 20 percent for the test set. Deep learning models, on the other hand, often use something closer to a 90/10 split to expand the training set. These are merely suggestions; the amount of available data decides the issue. If the available data is plentiful, try 90/10;

otherwise, maybe 80/20 to ensure that there is adequate diversity in the test set to convince yourself that the model has learned. A 90/10 split implies a training set and a held-out test set. It also implies selecting a model architecture and associated learning parameters in advance. What I mean by "learning parameters" (hyperparameters) will be come clear in time.

What if the selected architecture isn't adequate? Perhaps the model is too small, with too few weights and biases to capture the necessary mapping from inputs to desired outputs. Or, it's too large, and insufficient training data exists to condition the model. One way to detect such is to watch the training process, which is something we'll do in the projects, but another is to use a *validation set*, a training set used not to condition the model but to assess the training process to decide whether the model is learning correctly or whether the hyperparameters are appropriate. Exactly how the validation set factors into training the model is the subject of Steps 5 and 6 of The Process; however, the critical point is that the validation set is used to assess the current state of the model during training, but not to update the weights and biases of the model. Think of the validation set as a middle ground between what conditions the model parameters and what evaluates the final model.

It's tempting to want to assess the model during training with the test set and forgo the validation set, thereby making the training set all the bigger. However, doing so will bias the model in favor of the test set since we're using test set performance to decide how to train the model. If we then claim performance based on the test set is a fair assessment of the model, we'll be fooling ourselves to some degree. It's akin to cherry-picking data for presentation. Using the validation set to control model construction while reserving the test set for final assessment is the fairest approach, assuming both are appropriately selected from the original pool of possible training data.

Let's get practical with an example. We have 100,000 labeled samples, pairs (x_i, y_i) for $i = 0, 1, ..., 99999$ with x an image of a bird and y an integer indicating the class. We want to create three disjoint sets – training, validation and testing – using an 80/10/10 split. How do we do it so that we feel confident that each set appropriately represents the actual mix of classes?

We have options. If we feel pedantic, we might separate the pool of data by class, then put 80 percent of each class, randomly selected, into the training set, another 10 percent in the validation set, and the remaining 10 percent in the test set. This approach ensures that the mix of classes between the three is balanced and follows the true balance (the prior probabilities).

However, if the data pool is sufficiently large, we might rely instead on random sampling. What "sufficiently large" means is up to you, but 100,000 samples seems large enough. In that case, we scramble the order of the dataset to mix things (remember the iris flowers of Chapter 1 came to us in class order). Then, create the three datasets, passing the first 80 percent to the training set; the next 10 percent for validation and the remainder for testing. On average, such a mix will be close enough to the original blend of classes per dataset that we'll likely have no difficulty down the line.

A caveat is required. If one or more of the classes are rare, we are better off being pedantic to ensure that the smaller validation and test sets have adequate numbers of examples of rare classes to assess the model meaningfully.

To recap: we must have train and test sets, and a validation set if we are concerned that we have not selected an adequate architecture or hyperparameters. In practice, we build intuition with experience and will learn how to assess a training process by watching the behavior of the loss function, thereby making the validation set less necessary.

2.3.2 Data Augmentation

We seldom have enough training data. Therefore, most of the time, we want to enhance our existing datasets, now split between train, validation and test, by *data augmentation*. Data augmentation takes each real training sample and alters it slightly so that it can be interpreted as a new sample of the class. For images, this alteration typically involves small rotations, flips, shifts, scaling and passing the pixels through a function to alter the color or intensity (gamma correction). The critical point is that the augmented image remains a plausible member of the original image class. Clearly, rotations, flips, scaling, and color adjustments fit the bill.

Figure 2.4 illustrates the augmentations we'll employ in the projects. The original singing Western Meadowlark is in the upper left, along with five augmented versions (top row): flip left-right and gamma correction, (bottom row) 5-degree rotation, scale to 120 percent and shift in x and y. In all cases, the resulting image was center-cropped to keep a uniform size without artifacts like blank spaces when rotating.

There is really no reason why individual augmentations cannot be cumulative. An augmentation routine might accept the original image as input and then apply augmentations successively if a randomly selected value is below a given threshold. The output image is the result of a randomly selected set of augmentations, thereby expanding the range of images available to the training set.

Augmentation enhances model generalization by better representing the parent distribution, in other words, by illustrating likely real-world variation, which aids the model by encouraging it to ignore minute details inherent in the training set in favor of general features of the classes. Additionally, augmentation increases the training set's size, which is related to the number of parameters the model can successfully condition. A larger model requires, in general, a larger training set.

Augmentation is also valuable when we have imbalanced data. If a particular class is rare relative to the others, augmentation of that class might help the model learn it well nonetheless.

Figure 2.4 Image augmentations (left to right, top to bottom): original, flip left-right, gamma correction, rotate 5 degrees, scale 120 percent, shift in x and y.

AI people might talk about data augmentation as a *regularizer*. A regularizer encourages models to learn the classes' true distinguishing characteristics without focusing on the minutiae of the dataset used to train the model. For example, a regularizer, like an augmented dataset, might help the model ignore the fact that many of the images for a particular class were taken in one lighting condition or at a specific time of year. Without regularization, the model might begin to perceive (beware anthropomorphizing!) the fact that the trees in the images lack leaves as a true characteristic leading to the presence of the bird. Of course, it might also be the case that the presence of the bird is strongly associated with autumn. Context matters.

2.4   Architecture Selection and Training

Step 5: Select a model architecture and associated hyperparameters
Step 6: Train the selected model using the training set

Selecting a model architecture, associated hyperparameters, and then training on a dataset is too vast a topic, to say nothing of an active research area, to cover entirely here. Instead, I'll focus on the three architectures we'll explore in the projects, architectures suitable for images, and the relatively small datasets we will work with. This section introduces the architectures, including some new layer types, and then walks through a training exercise using a standard machine learning dataset.

Chapter 3 is where we configure our development environment, so my expectation for this chapter is that you read through the description. I'll show you some code and point you to the code for this chapter on the book's GitHub site, but I don't expect you to run it just yet. Feel free to return to this section after completing Chapter 3.

2.4.1   Model Architectures

Models must be realized in code; the best way is via an existing deep-learning toolkit. Fortunately, there are several to choose from, with two leading the pack: PyTorch and TensorFlow/Keras. Learning both is a good idea for those seeking to make AI something more than a hobby. We'll stick with TensorFlow from Google and the Keras Python framework that runs on top of it. Both leading toolkits are open source.

We'll focus on three CNN architectures: LeNet-5, VGG8, and ResNet-18. Practical CNN architectures use the convolutional, pooling and dense layers we discussed in Chapter 1 along with three new layers: batch normalization, dropout and global average pooling. Global average pooling is used by the ResNet-18 architecture, while LeNet-5 and VGG8 use dropout. All three architectures use batch normalization. Let's dive into the architectures beginning with LeNet-5. I implemented all three for you in Python. I recommend reviewing *lenet5.py*, *vgg8.py* and *resnet18.py* before proceeding.

LeNet-5

The original CNN architecture came out of researcher Yann LeCun's lab in the late 1990s. The published architecture has come to be known as "LeNet-5" and is shown schematically in Figure 2.5. The corresponding Keras code is in Listing 2.1.

The listing defines a function, LeNet5, that accepts a tuple describing the shape of an input tensor (input_shape) and, optionally, the number of unique classes in the

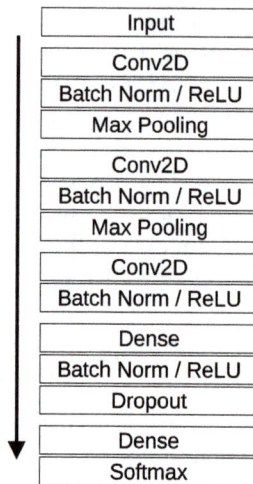

Figure 2.5 The LeNet-5 architecture

```
def LeNet5(input_shape, num_classes=10):
    inp = Input(input_shape)
    _ = Conv2D(6, (3,3))(inp)
    _ = BatchNormalization()(_)
    _ = ReLU()(_)
    _ = MaxPooling2D((2,2))(_)
    _ = Conv2D(16, (3,3))(_)
    _ = BatchNormalization()(_)
    _ = ReLU()(_)
    _ = MaxPooling2D((2,2))(_)
    _ = Conv2D(120, (3,3))(_)
    _ = BatchNormalization()(_)
    _ = ReLU()(_)
    _ = Flatten()(_)
    _ = Dense(84)(_)
    _ = BatchNormalization()(_)
    _ = ReLU()(_)
    _ = Dropout(0.5)(_)
    _ = Dense(num_classes)(_)
    outp = Softmax()(_)
    return Model(inputs=inp, outputs=outp)
```

Listing 2.1 Building a LeNet-5 model with Keras

dataset (default 10). The return value is a Model object. The input shape is $H \times W \times C$ given as (H, W, C) for image height, width, and number of channels (1 for grayscale, 3 for RGB).

I suspect you've noticed the excessive number of underscore characters (_). Keras builds models layer by layer, mimicking the flow of Figure 2.5. The output of a layer is passed as the input to the next layer. In most cases, there is no need to reference earlier layers beyond the input to the next layer, so dummy variables are often used and reused. Many developers prefer underscores for historical reasons, hence their appearance here. Other developers use x, but the variable name is irrelevant.

The first layer is `Input`. It accepts `input_shape` and returns a Keras `Input` object assigned to `inp`. We need to refer to the input layer when forming the model, so we store the layer in `inp` instead of underscore. Next comes a two-dimensional convolutional layer:

```
_ = Conv2D(6, (3,3))(inp)
```

The layer accepts the previous layer's output, `inp`, as input, along with parameters for the convolutional layer. In this case, the returned Keras convolutional layer is placed in underscore.

The syntax might be unfamiliar. Keras is an object-oriented toolkit, meaning `Conv2D` is the name of a class. The arguments, 6 for six filters and (3,3) for 3×3 kernels in the filters, are passed to the constructor for class `Conv2D`. The object instance is returned and immediately used with the overloaded function call syntax (`__call__` method) to tell the convolutional layer that its input is `inp`, the output of `Input`. The remainder of the model is built, layer by layer, using this approach.

Layers are often grouped to form repeated blocks. LeNet-5 uses three blocks of

```
_ = Conv2D(n, (3,3))(_)
_ = BatchNormalization()(_)
_ = ReLU()(_)
_ = MaxPooling2D((2,2))(_)
```

for n filters per block. The last block replaces `MaxPooling2D` with `Flatten`. Collectively, these layers form the portion of the network that transforms the input image into a representation (feature vector) that the remaining layers use to assign a class label.

We know what the `ReLU` and `MaxPooling2D` layers do based on their names. The `ReLU` layer applies the rectified linear unit activation function to every element of the input tensor. Recall that this function leaves positive values alone and replaces negative values with zero. The shape of the input tensor is not altered by the `ReLU` layer.

The `MaxPooling2D` layer uses 2×2 max pooling to spatially reduce the input tensor dimensions by a factor of two keeping the largest value in each 2×2 block. If the input tensor is of shape (64,64,3), then the output tensor is of shape (32,32,3) where the number of channels remains unchanged.

Batch normalization, like `ReLU`, does not alter the dimensions of its input. Instead, it first adjusts the input tensor to have zero mean and standard deviation one, the normalization part, then applies a learned scaling value and offset to produce the output tensor. Mathematically,

$$\hat{x} = \frac{x - \mu}{\sqrt{\sigma^2 + \epsilon}}$$

$$y = \gamma \hat{x} + \beta$$

for input tensor x of shape $H \times W \times C$, output tensor y of the same shape and C-dimensional vectors μ, σ, γ and β. Here, ϵ is a small scalar value to avoid division by zero problems

Input
ConvBlock 64
ConvBlock 128
ConvBlock 256
Flatten
DenseBlock 2048
DenseBlock 2048
Dense
Softmax

ConvBlock:
Conv2D
Batch Norm / ReLU
Conv2D
Batch Norm / ReLU
Max Pooling

DenseBlock:
Dense
Batch Norm / ReLU
Dropout

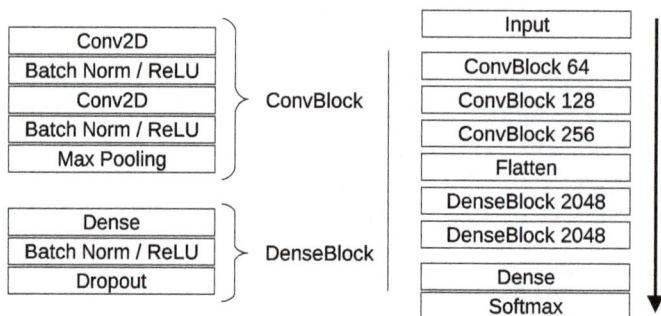

Figure 2.6 The VGG8 architecture

($\epsilon \approx 10^{-7}$). It's essential to note that the C-dimensional vectors are applied by channel to each of the $H \times W$ values in that channel. In other words, for channel i, subtract a single scalar value, μ_i, from each of the $H \times W$ values, etc. A strict understanding of how batch normalization is applied is not critical; remember that it scales an input tensor to make its values suitable for the remainder of the network.

Only Dropout remains. Dropout is a training trick that randomly sets values in the input tensor to zero according to the given probability, here 0.5. During training, the Dropout layer will flip a coin for each value in the input tensor. If heads, set the value in the output tensor to zero. If tails, retain the original value. Dropout helps, in many cases, by simulating an ensemble of many thousands of subnetworks using the random zeroing of values to encourage the weights in each layer to become less correlated and more responsive to actual data values. Dropout acts like a regularizer to make it more likely that the model will learn meaningful aspects of the data useful at inference time to correctly assign class labels without focusing on specific details in the training data. Dropout is typically applied to the output of the dense layers.

LeNet-5 is a "small" network with 181,114 trainable parameters and is best used with images only a few tens of pixels on a side.

VGG8

The VGG family of networks was developed by the Visual Geometry Group at the University of Oxford. They are a favorite of researchers seeking to transform images into embedding vectors. For us, the VGG8 version with eight convolutional and dense layers serves as a workhorse because it's well suited to the datasets used by the projects. Figure 2.6 presents the network's architecture as a series of repeated blocks. Think of VGG8 as an expanded version of LeNet-5. In code, VGG8 becomes Listing 2.2.

```
def ConvBlock(_, filters):
    _ = Conv2D(filters, (3,3), padding='same')(_)
    _ = BatchNormalization()(_)
    _ = ReLU()(_)
    _ = Conv2D(filters, (3,3), padding='same')(_)
    _ = BatchNormalization()(_)
    _ = ReLU()(_)
    return MaxPooling2D((2,2))(_)
```

```
def DenseBlock(_, nodes):
    _ = Dense(nodes)(_)
    _ = BatchNormalization()(_)
    _ = ReLU()(_)
    return Dropout(0.5)(_)

def VGG8(input_shape, num_classes=10):
    inp = Input(input_shape)
    _ = ConvBlock(inp, 64)
    _ = ConvBlock(_, 128)
    _ = ConvBlock(_, 256)
    _ = Flatten()(_)
    _ = DenseBlock(_, 2048)
    _ = DenseBlock(_, 2048)
    _ = Dense(num_classes)(_)
    outp = Softmax()(_)
    return Model(inputs=inp, outputs=outp)
```

Listing 2.2 The VGG8 architecture

The VGG8 function builds the model by repeated calls to ConvBlock and DenseBlock. Notice that the Conv2D layers use padding='same'. Doing so zero pads the input tensor, causing the convolution operation to produce a same-sized output in height and width. If the input tensor is of dimensionality $H \times W \times C$ and the Conv2D layer is requested to produce Y filters, the output is then an $H \times W \times Y$ tensor.

The VGG8 architecture is clearly divided into two parts. The first uses convolutional and pooling layers to transform the input image into a new representation, which becomes the input to the dense layers when flattened into a vector. The dense layers act like a traditional MLP to process the new representation and ultimately decide on class membership via the index of the largest value in the output softmax vector. Notice that the number of convolutional block filters increases with network depth from 64 to 256. This trend is common and helps the model learn abstract representations of the input image that prove useful as a feature vector for the top-level dense layers.

With nearly 14 million trainable parameters (13,762,890, to be exact), VGG8 is the biggest model we'll use in the projects. That said, my experience is that it produces reasonable results with even modest-sized training sets.

ResNet-18

The residual network family, known as ResNets, are the go-to architectures for computer vision tasks. Residual networks ask the model to learn offsets added to the input to a series of layers, the residual block. The advantage gained by doing so allows for truly deep neural networks with dozens to over one hundred layers. We'll restrain ourselves to the simplest of the standard ResNets, the ResNet-18 architecture with 18 trainable layers, as shown schematically in Figure 2.7.

The figure may appear daunting at first, but closer inspection reveals that it is nothing more than a collection of repeated residual blocks, each learning an increasing number of

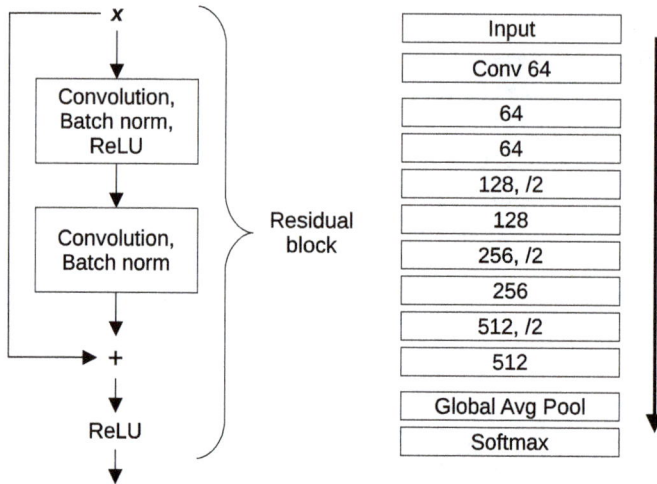

Figure 2.7 The ResNet-18 architecture

filters, culminating in a new layer, the global average pooling layer. The global average pooling layer replaces the dense layers of VGG8 and LetNet-5. The "/2" label indicates convolution with a stride of 2 in each spatial dimension, effectively reducing the spatial dimensions ($H \times W$) by two and mimicking the pooling layers found in VGG8 and LeNet-5.

In code, the ResNet-18 model becomes Listing 2.3.

```
def ResidualBlock(x, filters, downsample=False):
    if (downsample):
        strides = (2,2)
        inp=Conv2D(filters, (1,1), strides=strides,
                   padding='same')(x)
    else:
        strides = (1,1)
        inp = x
    _ = Conv2D(filters,(3,3),strides=strides,padding='same')(x)
    _ = BatchNormalization()(_)
    _ = ReLU()(_)
    _ = Conv2D(filters,(3,3),strides=(1,1),padding='same')(_)
    _ = BatchNormalization()(_)
    _ = Add()([_, inp])
    return ReLU()(_)

def ResNet18(input_shape, num_classes=10):
    inp = Input(input_shape)
    _ = Conv2D(64, (3,3), strides=(1,1), padding='same')(inp)
    _ = ResidualBlock(_, 64, downsample=False)
    _ = ResidualBlock(_, 64, downsample=False)
    _ = ResidualBlock(_, 128, downsample=True)
    _ = ResidualBlock(_, 128, downsample=False)
```

```
   _ = ResidualBlock(_, 256, downsample=True)
   _ = ResidualBlock(_, 256, downsample=False)
   _ = ResidualBlock(_, 512, downsample=True)
   _ = ResidualBlock(_, 512, downsample=False)
   _ = GlobalAveragePooling2D()(_)
   _ = Dense(num_classes)(_)
outp = Softmax()(_)
return Model(inputs=inp, outputs=outp)
```

Listing 2.3 The ResNet-18 architecture

A first convolutional layer performs an initial pass over the input image with same padding to produce a $H \times W \times 64$ tensor that we can interpret as the model's first transformation from an image to a new representation containing essential information about edges, textures, and colors.

Next comes eight residual blocks, each increasing the number of learned filters from 64 to 512. The downsample keyword indicates whether the block uses a 1×1 or 2×2 stride over the input tensor. Convolving with a 2×2 stride reduces the spatial dimensions of the tensor by a factor of two, like the max pooling layers in VGG8 and LeNet-5. The final step in each residual block adds the tensor produced by the block to the block's input. In that way, the block's goal is to produce an offset from the input tensor instead of an entirely new tensor. Learning an offset tensor for each residual block gives the network the flexibility to learn an adjustment, even if the proper adjustment for that layer is to do nothing (an offset of zeros).

The residual blocks ultimately produce an output tensor with 512 channels. The GlobalAveragePooling2D layer acts like a max pooling layer but averages over the spatial dimensions across all channels. If the input to the global average pooling layer is $H \times W \times 512$, the layer averages the $H \times W$ values in each channel to produce a 512-element output vector. Passing this 512-element vector, the learned new representation of the input image, to the final Dense and Softmax layers results in the model's output vector indicating its "belief" as to the correct class label for the input.

The ResNet-18 model's 11 million trainable parameters make it nearly the size of VGG8. For huge training sets, ResNet models tend to out perform VGG-style models. In practice, however, especially for the size datasets typically encountered when applying deep learning in the real world, my experience has been that VGG8 usually produces slightly better models. Your mileage may vary, and for any task, it is critical to test multiple architectures if time and resources allow.

2.4.2 Concerning Hyperparameters

Hyperparameters control the training process. Recall that training involves initializing the network using intelligently selected random values, which we envision as a point in a high-dimensional space, the space of all the weights and biases in the network. From there, gradient descent and backpropagation are used to "move" to a better place in the parameter space that minimizes the network's errors on the training data. The final position in parameter space is, coordinate by coordinate, the set of weights and biases for the now-trained network.

Gradient descent involves two steps to move from the current position in parameter space to the next: knowledge of how a slight change in each parameter affects the network's overall error and how big a step to take in the direction that reduces the error. In practice, using all the training data for each gradient descent step is infeasible. Instead, a random subset of the training data is used to estimate the network's error before updating the parameters. Using all of the training data to take a gradient descent step is known as *batch training*; therefore, using a subset is known as *minibatch training*. A key hyperparameter is the number of training samples in a minibatch, typically a power of two, to play nicely with hardware. Estimating the network's error with its current set of weights and biases by a minibatch of samples introduces randomness to the training process, hence referring to minibatch training as *stochastic gradient descent*. Stochastic gradient descent is faster and has the added benefit of sometimes moving in a direction through the error (or loss) landscape that isn't ideal, but might help the model avoid a local minimum and arrive at a better place in the end. An *epoch* implies passing all the training data through the network. If there are N samples in the training set, an epoch uses each of the N samples once before using any of the training data a second time. If the minibatch is M samples, then the number of minibatches per epoch is

$$\text{minibatches per epoch} = \frac{\text{training set samples}}{\text{minibatch size}} = \frac{N}{M}$$

implying that many gradient descent steps (weight and biase updates) per epoch. The minibatch size and number of training epochs are perhaps the two most essential hyperparameters and the only two we'll concern ourselves with for the projects.

Taking a gradient descent step implies a step size. Backpropagation tells the training algorithm how much a slight change in the current value of a weight affects the network's error, with all such values forming a direction in the parameter space (a gradient). However, the gradient does not specify how far to step in that direction. For that, we need a step size known as the *learning rate*. Classic stochastic gradient descent (SGD) uses the gradient (a set of partial derivatives) along with the step size, to alter each weight and bias:

$$w_i \leftarrow w_i - \eta \frac{\partial L}{\partial w_i}$$

for weight i of the network, step size η and error (loss) L over the current minibatch. Additional terms are often present, like momentum to retain a portion of the previous weight value and weight decay to help penalize weights that get too large.

Fortunately for us, however, classic SGD is often replaced by more advanced optimization algorithms that automatically manage most of the parameters involved. In practice, we seldom need to bother adjusting advanced optimization algorithm parameters, meaning we will be successful using them out of the box. For example, most of our projects will make due with the Adam optimizer, one widely used by the deep learning community and one well suited to training computer vision models for classification.

Let's finish this highly theoretical section with an example that puts everything together to make the training process concrete.

2.4.3 A Training Exercise

Let's build and train a model. The dataset for this section is a standard machine-learning workhorse known as CIFAR-10. It consists of 50,000 32×32 RGB images for training and another 10,000 for testing. There are ten categories: airplane (0), automobile (1), bird (2), cat (3), deer (4), dog (5), frog (6), horse (7), ship (8) and truck (9). One of the categories is birds, so I feel justified in using the dataset.

Building CIFAR-10

The dataset doesn't come to us prepackaged. Well, it does in various toolkits, but we're ignoring that fact here to gain experience in constructing and preprocessing data. First, download the raw binary data directly from the CIFAR-10 website:

```
https://www.cs.toronto.edu/~kriz/cifar-10-binary.tar.gz
```

and expand the file like so:

```
> tar xzf cifar-10-binary.tar.gz
```

where, as always, > is the command line prompt character, whatever it happens to be for your system.

The raw data is packaged in bundles of images and stored as unsigned integers in a nonstandard format. I already wrote a bit of Python code for you to extract the image data and store it in as NumPy arrays. You'll find the code in *build_cifar10.py*. Run the code like so:

```
> python3 build_cifar10.py 8675309
```

When finished, you'll be the proud owner of four new disk files:

```
cifar10_xtrain.npy
cifar10_ytrain.npy
cifar10_xtest.npy
cifar10_ytest.npy
```

The files use a naming convention we'll maintain throughout the book where xtrain refers to the training set images and ytrain the corresponding labels, with the test set versions following.

Let's run Python and load the training files to understand how they are packaged:

```
> python3
...
>>> import numpy as np
>>> x,y = np.load("cifar10_xtrain.npy"), np.load("
    cifar10_ytrain.npy")
>>> x.shape, y.shape
((50000, 32, 32, 3), (50000,))
```

The images (x) are stored as a four-dimensional array. Imagine a stack of RGB images where the first index selects a 32×32×3 array representing a single image, 3 channels each for red, green, and blue. The y array, a one-dimensional vector, contains the corresponding image label:

```
>>> x[111].shape
(32, 32, 3)
>>> y[111]
8
```

The 111-th (eleventy first) image is a 32×32×3 array, as expected, and, according to y[111] is of a ship. If you have Pillow installed, these commands will convert the image from a NumPy array to a PIL array, then display it

```
>>> from PIL import Image
>>> Image.fromarray(x[111]).show()
```

to convince you that the image is indeed of a ship.

I recommend reading through *build_cifar10.py* to understand the general process. The first part reads the mysterious 8675309 that I asked you to include when running the code. This number serves as a pseudorandom number seed to ensure that you build the same version of the dataset I did. The second part of the code plows through the raw image batches to fill in two NumPy arrays, x and y, as shown in Listing 2.4.

```
if (len(sys.argv) > 1):
    np.random.seed(int(sys.argv[1]))
x,y = [],[]
for batch in range(1,6):
    d = np.fromfile('cifar-10-batches-bin/data_batch_%d.bin' %
                    batch, dtype="uint8")
    for k in range(10000):
        off = k*3073
        label = d[off]
        off += 1
        im = np.zeros((32,32,3), dtype="uint8")
        im[:,:,0] = d[(off+ 0):(off+ 0+1024)]
                    .reshape((32,32))
        im[:,:,1] = d[(off+1024):(off+1024+1024)]
                    .reshape((32,32))
        im[:,:,2] = d[(off+2048):(off+1024+2048)]
                    .reshape((32,32))
        x.append(im)
        y.append(label)
x,y = np.array(x), np.array(y)
```

Listing 2.4 Processing the raw CIFAR-10 image data

Listing 2.4 is tailored to the peculiarities of how the raw data is stored. However, we should make no assumption about how the data is stored, so next we permute the order of the images (and associated labels), then store the results on disk:

```
idx = np.argsort(np.random.random(len(x)))
x,y = x[idx], y[idx]
np.save("cifar10_xtrain.npy", x)
np.save("cifar10_ytrain.npy", y)
```

The first two lines are worth remembering: we'll see similar lines of code throughout the projects to scramble data ordering. The first line defines a vector, idx, of the same length as x, containing a random permutation of the numbers $[0, N - 1]$ where N is the length of x. The call to random generates that many pseudorandom floating-point values which are then passed to argsort to return the order of indices that would sort the values. The second line indexes x and y with idx to shuffle the values in sync so images and labels remain properly associated.

Scrambling the order, especially if splitting a single collection of images into separate train and test sets, is essential to roughly balance the relative frequency of each class within the datasets. CIFAR-10 is split into train and test for us, but scrambling the order is still a good idea because it has the same effect when training with minibatches (more on that momentarily). Finally, the remainder of the code repeats the process to build the test dataset.

Building the LeNet-5 Model

We have train and test datasets; we're ready to build the model in code. The three models we'll use in the projects are contained within *lenet5.py*, *vgg8.py*, and *resnet18.py*. The projects will import the model code from these files, but if we execute the files directly, we'll instead build and train models on the CIFAR-10 datasets just created.

```
batch_size = int(sys.argv[1])
epochs = int(sys.argv[2])
outdir = sys.argv[3]
nsamp = 50000 if len(sys.argv) < 5 else int(sys.argv[4])
num_classes = 10
img_rows, img_cols = 32, 32
input_shape = (img_rows, img_cols, 3)

x_train = np.load("cifar10_xtrain.npy")[:nsamp]
ytrain = np.load("cifar10_ytrain.npy")[:nsamp]
x_test = np.load("cifar10_xtest.npy")
ytest = np.load("cifar10_ytest.npy")

x_train = x_train.astype('float32') / 255
x_test = x_test.astype('float32') / 255

y_train = keras.utils.to_categorical(ytrain, num_classes)
y_test = keras.utils.to_categorical(ytest, num_classes)
```

```
N = int(0.1*len(x_train))
x_val, x_train = x_train[:N], x_train[N:]
y_val, y_train = y_train[:N], y_train[N:]
```

Listing 2.5 Loading and preprocessing the datasets

Listing 2.5 loads and preprocesses the CIFAR-10 train and test data. We'll use a similar process in the projects. First, command line arguments are parsed, including setting nsamp appropriately. Next, essential variables are defined to specify the number of classes (10), and the shape of each input, a $32 \times 32 \times 3$ array.

The datasets themselves are loaded next. We'll work with datasets that fit in memory and are stored as NumPy arrays. As the are, however, the data is not in a ready-to-use format. The image data must be normalized from [0, 255] to [0, 1] by dividing by 255. More cryptic are the lines applying the Keras to_categorical utility function to the labels.

Labels are stored as integers, [0, 9], but the model has 10 outputs, a softmax likelihood vector, where each element represents the model's belief that the input is of that particular class. Therefore, the integer label must be transformed into a 10-element one-hot vector of zeros with a single one at the position matching the label. For example, if the label is 4, the corresponding one-hot vector becomes

```
0 0 0 0 1 0 0 0 0 0
```

for direct comparison with the softmax output vector. The final block of code in Listing 2.5 sets 10 percent of the training data aside for validation.

```
model = LeNet5(input_shape)

model.compile(loss=keras.losses.categorical_crossentropy,
              optimizer=keras.optimizers.Adam(),
              metrics=['accuracy'])

history = model.fit(x_train, y_train,
              batch_size=batch_size,
              epochs=epochs,
              verbose=1,
              validation_data=(x_val, y_val))

pred = model.predict(x_test, verbose=0)
plabel = np.argmax(pred, axis=1)
cm, acc = ConfusionMatrix(plabel, ytest)
mcc = matthews_corrcoef(ytest, plabel)
txt = 'Test set accuracy: %0.4f, MCC: %0.4f' % (acc,mcc)
print(cm)
print(txt)
```

Listing 2.6 Defining, training and testing the LeNet-5 model

Data loaded and preprocessed, we're now ready to define, train and test the model as in Listing 2.6. A call to `LeNet` returns the Keras model. Before training the model, however, we must "compile" it to specify learning parameters like the loss function, here Keras' standard cross-entropy loss that interprets the output as a probability distribution over classes. We also specify the Adam optimizer using its default parameters, and finally tell Keras that it should track accuracy along with the loss during training (`metrics`).

Training itself becomes a call to the model's `fit` function. The training data and one-hot labels are supplied, along with the minibatch size and number of epochs. Setting `verbose=1` produces output, allowing us to track the training process. The validation data is used here as well. The `fit` method returns history information tracking metrics over the training process. Code at the end of *lenet5.py*, not shown, uses this information to produce an output plot.

The remainder of the code in Listing 2.6 uses the now-trained model to make predictions on the held-out test data (`x_test`). The output of `predict` is a $M \times N$ matrix of softmax values for M test set samples and N classes. The index of the largest `pred` column for any row is the label we should assign to the sample. This is what `np.argmax` gives us in `plabel`.

From `plabel` and the known test set labels in `ytest` we get the confusion matrix via a call to `ConfusionMatrix` (top of *lenet5.py*) and the MCC courtesy of an imported Scikit-Learn function.

Training the LeNet-5 Model

Executing *lenet5.py* without command line arguments produces:

```
lenet5 <minibatch> <epochs> <outdir> [<nsamp>]

    <minibatch> - minibatch size (e.g. 64)
    <epochs>    - number of training epochs (e.g. 16)
    <outdir>    - output directory name
    <nsamp>     - number of training samples
                  (optional, all if not given)
```

telling us what the code expects on the command line. Note that your system might also produce warning messages about CPUs and instructions. These are from the TensorFlow library and may be safely ignored (my system produces such messages, for example).

The code expects three command line arguments, `minibatch`, `epochs` and `outdir`, along with an optional fourth, `nsamp`. The first two specify the number of training samples in each minibatch and the number of epochs over which to train. The third argument is the name of an output directory and the fourth, if given, specifies the number of training samples to use other than all 50,000.

Recall that the number of gradient descent steps used to update the weights in each epoch is the number of training samples divided by the minibatch size. Therefore, the total number of gradient descent steps used to train the model because we are ignoring any kind of early stopping is

$$\text{gradient descent steps} = \text{epochs} \times \frac{\text{training samples}}{\text{minibatch size}}$$

implying training with a minibatch of 128 and all samples for 24 epochs results in

$$\text{gradient descent steps} = 24 \times \frac{50,000}{128} = 9375 \text{ steps}$$

In other words, the randomly selected initial set of model weights and biases will be updated 9375 times before declaring the model "trained."

Let's train a LeNet model for 24 epochs using a minibatch size of 128. I used a command line like this one:

```
> python3 lenet5.py 128 24 results/lenet5_128_24
```

where the output directory is within the *results* directory that I previously created.

Training shows output from Keras, epoch by epoch:

```
Epoch 1/24
   loss: 1.6603 - accuracy: 0.4134 -
   val_loss: 1.6741 - val_accuracy: 0.3926
Epoch 2/24
   loss: 1.3075 - accuracy: 0.5354 -
   val_loss: 1.2121 - val_accuracy: 0.5750
Epoch 3/24
   loss: 1.1815 - accuracy: 0.5840 -
   val_loss: 1.1243 - val_accuracy: 0.6110
```

Here, loss refers to the model's current error on the training set; lower is better, while accuracy is the model's overall accuracy on the training data. Over time, this accuracy will approach 1.0, implying that the model is getting better and better at correctly classifying the training samples. Note that I edited the output to remove the progress bar shown when running the code. The progress bar follows each minibatch as it is passed through the model to update the weights.

That the training accuracy will ultimately approach 1.0 does not imply that the model is actually learning in a way that will make it useful when used in the wild. For that, we need to examine its performance on the validation set given per epoch as val_loss and val_accuracy.

We'll discuss the validation set more thoroughly in the next section, but for now, if the validation accuracy isn't increasing or hovers around random guessing, then we have a clue that the model isn't learning.

CIFAR-10 contains ten classes, meaning random guessing will produce an accuracy of about $1/10 = 0.1$ or 10 percent. After the first epoch, the model is already 39 percent accurate on the validation data, and this value only increases, so we have evidence that the model is learning something valuable about differentiating between classes.

Evaluating the Model

The LeNet-5 model trains quicking, even on my CPU-only test machine. After about 5 minutes, all 24 epochs are complete, and the code displays the confusion matrix produced when the model is run against the test set:

```
[[795  34  25  11  11   6   3   6  55  54]
 [ 26 842   3   8   1   1   2   1  25  91]
 [114  16 488  72 120  71  38  42  15  24]
 [ 49  20  63 448  78 170  49  57  29  37]
 [ 32  10  55  50 674  38  28  87  13  13]
 [ 31   9  52 161  45 572   9  92  12  17]
 [ 19  20  54  76  72  10 689  18  18  24]
 [ 29  10  27  26  69  47   5 741   5  41]
 [102  61   6  12   2   6   1   2 767  41]
 [ 45  96   5   9   4   5   3  10  33 790]]
Test set accuracy: 0.6806, MCC: 0.6458
```

Recall that a confusion matrix shows true labels across rows and model assigned labels down columns, numbering the class labels from 0 through 9 in this case. Perfection is a purely diagonal matrix. Clearly, the LeNet-5 model isn't perfect, but the diagonal elements (counts) are larger than the off-diagonal elements demonstrating some level of ability on the model's part. The test set's overall accuracy is 68 percent, nearly 7× random chance accuracy. The MCC value refers to the Matthews Correlation Coefficient, which is a more rigorous measure of model performance, especially if the classes are imbalanced (not the case here). The closer the MCC is to 1.0, the better the model.

A complete analysis of the model's performance would consider the off-diagonal elements of the confusion matrix to the point of examining the test set images, which leads to particular classification errors. Doing so might reveal cases where the model is struggling with certain types of inputs or classes and offer insight on how to improve the training set, perhaps by adding more of the subtypes the model finds difficult or hard negatives – negative examples similar to actual instances to give the model implicit guidance about how to learn to distinguish the two.

The output directory contains several useful files:

```
accuracy_mcc.txt
confusion_matrix.npy
error_plot.png
model.keras
```

The file *accuracy_mcc.txt* contains the "Test set accuracy" line displayed on the screen, while the file *confusion_matrix.npy* contains the printed confusion matrix as a NumPy array. The trained model itself is in *model.keras*. With this file, we can deploy the network for downstream tasks.

The last file, *error_plot.png* is shown in Figure 2.8. It presents the error on the training set and the validation set as a function of epoch. The training error decreases consistently, as expected from gradient descent. The validation error, however, is erratic, in part because there isn't that much of it (about 1000 samples) and because of the stochastic nature of the training process, especially when using an advanced optimizer like the Adam optimizer used here.

The validation error also plateaus at about 35 percent, meaning the model's prediction is wrong in nearly 2 out of every 5 samples. That the error plateaus and doesn't increase with epoch indicates that the model has learned, but also that the model has limited capacity to learn more related to what separates classes. The notion of "capacity" is

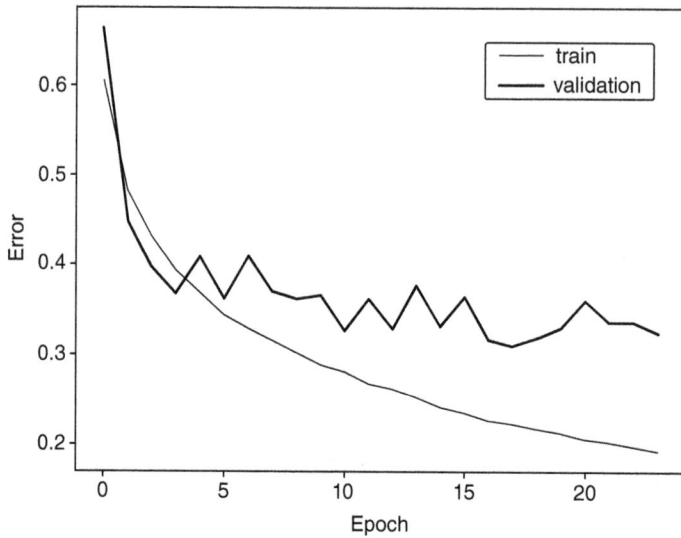

Figure 2.8 Training and validation error as a function of epoch.

slippery, but it is roughly affected by the number of training samples and the number of weights and biases in the network.

To run the same experiment with *vgg8.py* and *resnet18.py*, replace *lenet5.py*:

```
> python3 vgg8.py 128 24 results/vgg8_128_24
> python3 resnet18.py 128 24 results/resnet18_128_24
```

and then go get a cup of coffee or let it run overnight as the larger models take significantly longer to train: 143 minutes and 503 minutes on my test machine, respectively. The results, however, are improved relative to LeNet-5 with test set accuracies of 83.64 percent for VGG8 and 79.81 percent for ResNet-18. The stochastic nature of model initialization means your results will be different, but not significantly so.

2.5   Using the Validation Set

Step 7: Use the validation set, if any

The CIFAR-10 example in the previous section set training samples aside as validation data but did not use the data to decide when to stop training. Telling Keras to use validation data is a straightforward extension to the training process. The code we need is in *lenet5_early.py* and amounts to a modest modification of the original to go from

```
history = model.fit(x_train, y_train,
          batch_size=batch_size,
          epochs=epochs,
          verbose=1,
          validation_data=(x_val, y_val))
```

to

```
early = EarlyStopping(monitor='val_loss', patience=3,
                      restore_best_weights=True)

history = model.fit(x_train, y_train,
            batch_size=batch_size,
            epochs=epochs,
            verbose=1,
            validation_data=(x_val, y_val),
            callbacks=[early])
```

by defining an early stopping object (`early`) before telling Keras to use it during training via the `callbacks` keyword, which expects a list of callback functions to use after each training epoch. Setting the `patience` to 3 implies stopping training after 3 epochs where the validation loss does not improve. The best set of weights is then placed into the model prior to ending training.

My run of *lenet5_early.py* asked for 24 epochs, but the callback ended training after 14 to return a model that was 68.7 percent accurate on the test set.

It's essential to understand the philosophy behind the validation set in reference to the training and testing datasets. The training set is interpreted as a stand-in for the full, unknown parent distribution that can generate data of the kind we want a model to interpret. I think of the training set as a sample, an empirical representation of a continuous distribution.

The validation set is not used to learn model parameters, removing it from the sample serving as the estimate of the parent distribution but incorporating it as a driver for the training process for all the learning hyperparameters and architecture selection behind the final model. For example, if the current model architecture fails to do well on the validation set during training for the current set of learning parameters, this failure drives adjustments to the architecture and possibly the hyperparameters.

Only after the training and validation sets are used to produce the final model is the held-out test set provided as an independent assessment of the model's quality. We can't reasonably expect to assess using the training set because, by definition, gradient descent will eventually produce a good result. Philosophically, we shouldn't use the performance on the validation set, either, as we used the validation set to drive the training process. Only the pristine, held-out test set can give us a believable assessment of the model, provided it is of the same parent distribution and sufficiently representative of what the model will encounter when used.

2.6   Final Testing and Deployment

Step 8: Test the final model using the held-out test set
Step 9: Deploy the model for use in the wild

Training is complete. The next step is to use the held-out test set to evaluate the model before deployment in the wild. This section outlines each step.

2.6.1   Final Testing

The CIFAR-10 example calculated the confusion matrix, the overall accuracy and MCC, the latter via a Scikit-Learn function to handle multiclass problems.

The first thing one should do after passing the test set through the model is to use the known test set labels with the model's predicted class labels to calculate the confusion matrix. Examine the confusion matrix closely. Does it make sense? Does it seem consistent with the validation data accuracies given during training? Suppose the test set is from the same data-generating parent distribution as the train and validation data. In that case, the answer should be "yes." If it isn't, something is amiss with the dataset construction process, or the model didn't learn properly. Models that don't learn correctly often exhibit excessive preference for a few output classes while ignoring others.

The best evaluation tool for multiclass problems like CIFAR-10 is the confusion matrix. As we've learned, metrics like MCC can be calculated for multiclass problems, though this is generally best left to toolkit implementations.

This section describes a set of metrics for binary classification problems, those involving only two classes: a negative class (label 0) and a positive or target class (label 1). The general approach is to calculate the 2×2 confusion matrix, also known as a *contigency matrix*:

$$\text{confusion matrix} = \begin{bmatrix} \text{true negatives (TN)} & \text{false positives (FP)} \\ \text{false negatives (FN)} & \text{true positives (TP)} \end{bmatrix}$$

The 2×2 confusion matrix follows the structure of the multiclass matrix: diagonal elements (TN, TP) are correct classifications with off-diagonal elements (FN, FP) errors.

A good classifier will have low FN and FP counts. This much is intuitive. However, in practice, it is helpful to generate metrics from the confusion matrix, each based on the matrix elements themselves, or metrics calculated from them.

The file *metrics.py* implements a set of binary metrics. To use them, import:

```
>>> from metrics import ConfusionMatrix, Metrics
```

The function ConfusionMatrix is called first to calculate a confusion matrix from the test set results (predicted labels and true labels) passing that matrix to Metrics then returns a Python dictionary containing the calculated metrics.

Executing *metrics.py* trains a LeNet-5 model on the airplane and bird classes from CIFAR-10. For example,

```
> python3 metrics.py 64 16
```

trains for 16 epochs using a minibatch of 64. A run of the code produced the following output:

```
TN: 793    FP: 207
FN: 70     TP: 930

ACC: 0.86150    (accuracy)
MCC: 0.72988    (Matthews correlation coefficient)

TPR: 0.93000    (true positive rate, sensitivity, recall, hit
      rate)
TNR: 0.79300    (true negative rate, specificity)
```

```
FPR: 0.20700      (false positive rate)
FNR: 0.07000      (false negative rate)

PPV: 0.81794      (positive predictive value, precision)
NPV: 0.91889      (negative predictive value)

F1            : 0.87038
informedness  : 0.72300
markedness    : 0.73683
kappa         : 0.72300
```

There are 2,000 test set samples, half airplanes, and half birds. The confusion matrix is presented first with labels for each element. The code uses the given text as keys, e.g., 'TN', 'MCC', etc.

After the confusion matrix comes the accuracy (ACC) and MCC. Both approach 1.0 as the model's performance improves, and both should be considered first after examining the confusion matrix. In general, all metrics approach 1.0 as the model improves. Therefore, examining the confusion matrix, accuracy, and MCC is usually enough. Remember that imbalanced classes might skew the interpretation of the accuracy, so MCC is essential. The airplane versus bird model's accuracy was 86 percent with an MCC of 73 percent, both of which are indicative of a modestly performant model. The next set of metrics is what I call "second tier" because they are constructed from the confusion matrix. The text after the values identify the metric, which often goes by multiple names in the literature. For example, the *true positive rate (TPR)* is known, primarily in medical literature, as the "sensitivity," while the same value is the "recall" or "hit rate" in other contexts.

The rates can be interpreted as probabilities. The true positive rate (TPR) is the probability that a bird will be detected by the classifier, meaning 93 percent of the birds in the test set were assigned correctly to class 1. The *true negative rate (TNR)* (aka specificity) is the complement: the probability that an airplane (class 0) will be assigned by the model to class 0. Again, a well-performing model will make both of these rates close to 1.0.

The example model's high TPR implies success in detecting actual bird samples. The lower TNR, however, combined with a 21 percent FPR (false positive rate) tells us that even though the model is likely to capture actual bird samples, it is prone to calling airplanes birds, which shouldn't surprise us much if we think about it for a bit.

At times, the cost associated with a particular error makes ensuring the model acts one way or another more important. For example, if the model is screening for cancer, it is critical not to miss actual cancer, implying a high TPR (sensitivity). Alternatively, it might be acceptable for the model to have a higher false positive rate (FPR) and claim cancer detection when there is no cancer. Further testing from a positive result will reveal the mistake, while missing cancer may prove fatal. Thankfully, birding applications are likely to be of less import in their predictions. *Positive predictive value (PPV)* and *negative predictive value (NPV)* come next. The PPV is more often known as "precision" and it reflects the probability of the model being correct when it labels an input as belonging to class 1 (target). Note that this is not the same as the probability that a true instance of class 1 will be so labeled by the model (TPR). The airplane versus bird model's PPV of 82 percent is a measure of how much confidence we should place in the model's "bird" predictions. We expect it to be correct a little more than 4 out of 5 times in that case. The high negative predictive value, 92 percent, implies "airplane" predictions are correct better than 9 out of 10 times.

The final four metrics are for completeness. The *F1 score* is a metric constructed from the precision and recall (TPR):

$$F1 = 2 \times \frac{PPV \times TPR}{PPV + TPR} = \frac{2TP}{2TP + FP + FN}$$

F1 is the harmonic mean of the precision and true positive rate. When cast in terms of the confusion matrix, notice that F1 does not take the true negatives (TN) into account. Because of this, I tend not to favor reporting F1 for classification tasks such as those of the projects, as it tends to be somewhat on the optimistic side. That said, as with all the metrics, F1 will be near 1.0 if the model is highly accurate. *Informedness* and *markedness* have been put forward in the recent past as metrics that should be considered over the F1 score. Note, however, that the MCC is the geometric mean of the informedness and markedness:

$$MCC = \sqrt{\text{informedness} \times \text{markedness}}$$

implying that, once again, MCC captures the essential quality of the model.

The final metric is *Cohen's* κ (kappa). It is constructed from the observed accuracy (p_o) and the accuracy expected by chance (p_e):

$$\kappa = \frac{p_o - p_e}{1 - p_e}$$

with p_o and p_e derived from the confusion matrix. The Metrics function reveals the formulas. As we expect by now, $\kappa \to 1$ implies a better model. The code in *metrics.py* is sufficient for our experiments. Complete evaluation of binary models often incorporates the *Receiver Operating Characteristic (ROC)* curve, a graph representing how the model behaves over a range of *decision threshold* values. Typically, for a binary model, if the prediction likelihood exceeds the decision threshold of 0.5, we assign the input to class 1; otherwise, we label it class 0. The decision threshold affects the TN, FP, FN, and TP values in the confusion matrix and by extension the metrics based on the confusion matrix. If the threshold is varied smoothly, from, say, 0.1 through to 0.9 in fixed-sized increments, it becomes possible to calculate metrics at each threshold value. The ROC curve plots the true positive rate (TPR) as a function of the false positive rate (FPR) often labeled as $1 - TNR = 1 - $ specificity over the decsion thresholds.

The area under the ROC curve, *AUC*, is often given as an additional metric, though the primary utility of the curve lies in its visual appearance. The curve makes it possible to assess how the model behaves when altering the decision threshold to optimize performance for different situations. For example, by preferring to accept more false positives to ensure better capture of actual target samples (the cancer detection scenario).

However, at this point in the book, it seems unnecessary to spend time exploring the ROC curve or, perhaps even more useful, an MCC curve generated in the same manner. High-performing models will produce metrics near 1.0, so if our models return several metrics in that region, we'll consider the model good and move on.

2.6.2   Model Deployment

The model is complete and ready for deployment if the metrics applied to the test set are acceptable. Let's spend some time contemplating what that entails.

Deployment is task-specific, but a few general comments are possible. First, the model must be in a format supported by the software environment in which it will be used. This is obvious but includes paying attention to toolkit versions to ensure the model is supported correctly. In the best of all possible worlds, new software versions are entirely backward compatible. We don't live in the best of all possible worlds. Upgrading the toolkit mandates retesting the model and any code in which it is embedded.

Another point to remember is that any preprocessing applied to the model's inputs must be applied to inputs in the wild. For image-based models, this isn't much more than scaling to map the pixel values to [0, 1], but might become more involved as the model is used over time. Was the camera replaced? If so, ensure that it produces output suitable for the existing preprocessing routine. For example, if the old camera produces 10-bit grayscale images, the proper scaling factor is $2^{10} - 1 = 1023$, but a new camera producing 12-bit images requires a new scaling factor of $2^{12} - 1 = 4095$.

A long-term concern revolves around *data drift*, the subtle change in the statistical qualities of the input data over time due to changes in the environment (train for spring and it's now autumn) or in the sensors used (degradation of the camera via increased noise, etc.)

Data drift is most common in systems that run for a long time and can be challenging to detect. Periodic retesting of the model on new test sets derived from the system as it currently stands is helpful. If the model's performance suddenly decreases, something is wrong, or at least different, and retraining of the model might be required.

3. Configuring the Desktop Environment

The book's experiments require a particular desktop environment. Specifically, we need Python 3.X and a suite of Python libraries. This chapter introduces the main libraries with others installed on demand when necessary. The remainder of the chapter outlines how to configure different desktops: Linux, macOS, and Windows. I assume a recent version of Ubuntu Linux throughout the book. You should have no difficulty using macOS or Windows.

3.1 Introducing the Toolkits

Python is the primary language of deep learning, at least at the level we are pursuing here. Under the hood, Python relies on advanced libraries implemented in C++. This section presents the key libraries we'll use, beginning with NumPy, the library that makes Python useful for data science (ignoring Pandas, which rides on top of NumPy).

TensorFlow and Keras come next. They are intertwined and represent one of the two primary implementations of deep neural networks. The other is PyTorch, which you may wish to explore should you continue in this area. Both are equally capable of implementing any of the architectures you'll encounter in the literature, and we'll use PyTorch for the experiments involving CLIP embeddings (and we'll learn what that all means by the time we need it).

Matplotlib, Scikit-Learn, SciPy and Pillow (aka PIL) round out the core libraries. Matplotlib gives Python advanced graphing abilities based on the approach used by Matlab (hence the name). Scikit-Learn is a pure Python library for data science and classical machine learning. We'll use its facilities from time to time. Finally, SciPy adds higher-level scientific analysis tools to Python with NumPy as a base and Pillow enables image manipulation.

3.1.1 NumPy

Without NumPy, Python is largely unsuitable for scientific computing. Pure Python can be performant, but in the end, collections of data are ideally suited to array processing, which is what NumPy adds to Python along with an understanding of data types. A list of lists emulates a matrix, but for analysis of data at scale, Python alone isn't sufficient. NumPy is the foundation of the other toolkits discussed in this chapter. In Python, if A is a variable holding an integer, the expression A+1 adds one to that value. In NumPy, A might be an array of one, two, three or more dimensions. For example, a dataset of RGB images is usually represented in Keras as a four-dimensional array. Regardless of its dimensionality, the expression A+1 adds one to every element of the array.

A full review of NumPy's abilities is a book in itself. This book's GitHub site includes a *tutorial.pdf* providing a bare-bones introduction to NumPy (and Matplotlib, SciPy and Pillow). A few examples convey NumPy's flavor:

```
> python3
>>> import numpy as np    # NumPy is universally renamed 'np'
>>> A = np.array([[1,2,3],[4,5,6],[7,8,9]]) # list to matrix
>>> A
array([[1, 2, 3],
       [4, 5, 6],
       [7, 8, 9]])
>>> A+1
array([[ 2, 3,  4],
       [ 5, 6,  7],
       [ 8, 9, 10]])
>>> A * 10
array([[10, 20, 30],
       [40, 50, 60],
       [70, 80, 90]])
>>> B = np.array([11,22,33])        # a row vector
>>> B
array([11, 22, 33])
>>> C = B.reshape((3,1))            # a column vector
>>> C
array([[11],
       [22],
       [33]])
>>> A @ C                          # matrix multiplication
array([[154],
       [352],
       [550]])
>>> A[1,1] = 999                   # indexing is zero-based
>>> A
array([[ 1,   2, 3],
       [ 4, 999, 6],
       [ 7,   8, 9]])
>>> A @ A.T                        # A.T is transpose
array([[   14,   2020,   50],
       [ 2020, 998053, 8074],
       [   50,   8074,  194]])
```

NumPy broadcasts operators over arrays in a sensible fashion, which is why A+1 and A*10 alter each element (a matrix and scalar operation). The @ operator implements matrix multiplication; often historically implemented via np.dot as commonly encountered in deep learning code. Interpretation of the output from np.dot depends on the shape (dimensionality) of its inputs. Extensive documentation is available on the NumPy site (https://numpy.org).

3.1.2 TensorFlow and Keras

TensorFlow is a deep learning framework supported by Google. It is written in C++ and is one of the foundational toolkits for AI. Keras, by François Chollet, is a Python wrapper (and more) over TensorFlow. The two are bundled together – installing TensorFlow installs Keras. Keras is the backbone of the book's experiments.

As an example, consider a simple convolutional neural network implemented in Keras:

```
inp = Input(input_shape)
_   = Conv2D(6,  (3,3))(inp)
_   = ReLU()(_)
_   = MaxPooling2D((2,2))(_)
_   = Conv2D(16,  (3,3))(_)
_   = ReLU()(_)
_   = MaxPooling2D((2,2))(_)
_   = Conv2D(120,  (3,3))(_)
_   = ReLU()(_)
_   = Flatten()(_)
_   = Dense(84)(_)
_   = ReLU()(_)
_   = Dense(num_classes)(_)
outp = Softmax()(_)
model = Model(inputs=inp, outputs=outp)
```

Deep networks are constructed layer by layer, with the output of one layer often serving as the input to the next, thereby linking the different layers together to form the desired architecture. Here, the _ variable, allowed by Python, serves as the link.

In this example, three convolutional (Conv2D) and pooling (MaxPooling2D) layers are applied to an input image of dimensions input_shape to create a new feature representation vector, the output of Flatten. This vector is then passed through what amounts to a traditional neural network with a single hidden layer of 84 nodes. The final Dense and Softmax layers complete the network to produce an output vector of num_classes, where each element is the model's estimate of the input's class membership, [0, 1]. Typically, the label associated with this vector's maximum element is returned as the model's prediction. In this way, networks of arbitrary complexity can be constructed with the TensorFlow toolkit running beneath providing all the necessary implementation, including arranging for calculating parameter gradients (backpropagation) required by the stochastic gradient descent training process. Visit https://www.tensorflow.org/ for details.

3.1.3   Matplotlib

Matplotlib adds 2D and 3D plotting to Python. Every plot in the book was generated with Matplotlib and output either as a PNG file for viewing or an EPS (encapsulated PostScript) file for publication. Plotting works largely as expected. Consider:

```
import numpy as np
import matplotlib.pylab as plt

#  Matplotlib works with NumPy
x = np.linspace(0, 2*np.pi, 1000)
y = np.sin(x)

#  And builds plots piece by piece
plt.plot(x,y, linewidth=0.7, color='k')
plt.plot(x[::50], y[::50], marker='o',
         linestyle='none', fillstyle='none',
         color='k')

plt.xlabel("$x$")               # LaTeX aware
plt.ylabel("$\\sin(x)$")
plt.title("A sample plot")
plt.tight_layout()

plt.savefig("plot.eps", dpi=300)
plt.savefig("plot.png", dpi=300)
plt.show()
```

Figure 3.1 is the result. Everything from quick-and-dirty plots to complex publication quality plots with subplots are possible. Visit https://matplotlib.org/ for detailed information.

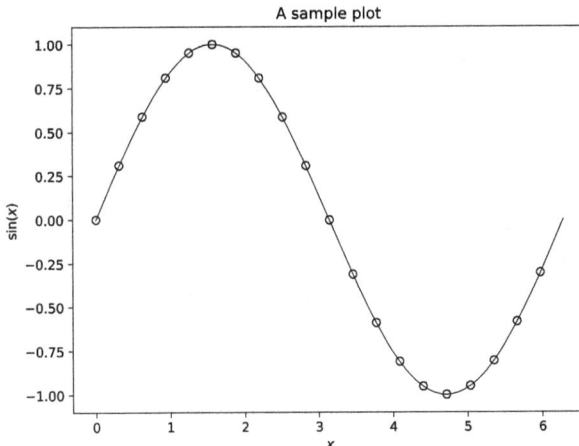

Figure 3.1 A sample Matplotlib plot

3.1.4 Scikit-Learn

TensorFlow and Keras are neural network-specific. Classical machine learning and a plethora of associated functionality is available courtesy of Scikit-Learn; see `https://scikit-learn.org/stable/`. I'll refer to Scikit-Learn as "sklearn" as that's how it's accessed within Python. Many experiments include `matthews_coffcoef` to calculate the Matthews correlation coefficient (MCC), perhaps the best single-value measure of a classifier's quality. Later chapters explore building top-level models on the embeddings produced by another model. The top-level model might be a traditional neural network or another classical machine learning model like *k*-nearest neighbors, random forest, or support vector machine, among others. The old-school approach is sometimes worthwhile, as we'll learn.

3.1.5 SciPy and Pillow

NumPy provides much, but not all, of what we require. In particular, NumPy lacks statistical analysis functions. For that, we need SciPy (`https://scipy.org/`). SciPy bills itself as "fundamental algorithms for scientific computing in Python" and that's as good of a description as any. I encourage you to explore what SciPy offers, we do it an injustice by treating it as a mere collection of a handful of stats functions.

That said, stats are easily calculated with SciPy. Consider,

```
> python3
>>> import numpy as np   # NumPy is always required
>>> from scipy.stats import ttest_ind, mannwhitneyu
>>> from scipy.stats import ttest_rel, wilcoxon
>>>
>>> A = np.random.normal(1, 1, size=10_000)
>>> B = np.random.normal(0.97, 2, size=10_000)
>>>
>>> t,p = ttest_ind(A,B)
>>> print("t=%0.4f, p=%0.5f" % (t,p))
t=2.1841, p=0.02897
>>> U,p = mannwhitneyu(A,B)
>>> print("U=%0.4f, p=%0.5f" % (U,p))
U=50842679.0000, p=0.03901
```

NumPy is required, then independent and paired tests are imported from SciPy, both parametric t-tests and nonparametric tests. I tend to use both during analysis rather than rely on normality assumptions.

Then, generate two sets of 10,000 samples from normal distributions; the first (A) has a mean and standard deviation of 1, while the second (B) has a slightly smaller mean of 0.97 and a wider standard deviation of 2.

The statistic (t or U) is displayed along with the p-value. In this case, a traditional threshold of 0.05 implies both tests provide evidence for rejecting the null hypothesis that the two samples are from the same distribution – though the evidence is weak and should be followed up with more observations, if possible.

This overview of the main toolkits is necessarily brief and incomplete. Working through the book enhances familiarity and drives home how to use the toolkits at the basic level we require. To dive deeper, consider the website tutorials or one of the many books on scientific programming with Python.

Readers familiar with R might benefit from learning about Pandas, a Python approach to statistics in the spirit of R. The analyses in this book are amenable to Pandas, if desired.

Let's configure our working environment. Jump to the operating system you are using, bearing in mind that the instructions will not handle every scenario.

3.2 Configuring Linux

I'm assuming Ubuntu 24.04 or later, though most full-featured Linux distributions will work just fine. The first step is to create an environment and then activate it and begin installing packages with pip3, Python's package installer:

```
> sudo apt-get install python3-venv  # if necessary
> python3 -m venv birds
> source birds/bin/activate
(birds) > pip3 install numpy
__--snip--__
```

The commands ensure that venv is present (if not already installed) before creating a new birds environment. Feel free to use any name you wish here. Next, source activates the environment, allowing us to install packages freely without interfering with existing environments or the versions used by the operating system. Finally, install numpy. I'm ignoring the installation messages here and in the following listing.

Assuming the above went well, we're now ready to install the remaining packages:

```
(birds) > pip3 install tensorflow-cpu
(birds) > pip3 install matplotlib
(birds) > pip3 install scikit-learn
(birds) > pip3 install scipy
(birds) > pip3 install pillow
(birds) > deactivate
```

The final deactivate command exits the environment. Use source again to activate before running the book's code examples. If there were no installation errors, you should be ready to jump to the next chapter.

3.3 Configuring macOS

Mac users should read the section above, as installation on macOS is nearly identical to Linux. When finished, come back here for the few differences. Ready? macOS uses brew in place of apt-get. Therefore, ensure that brew is installed:

```
/bin/bash -c "$(curl -fsSL
https://raw.githubusercontent.com/Homebrew/install/HEAD
    /install.sh)"
```

where the entire command should be on one line.

Then, the first set of installation steps becomes:

```
> brew install python@3.10
> python3.10 -m venv birds
> source birds/bin/activate
(birds) > pip3 install numpy
```

The `brew` command installs a version of Python 3.10, which works well with TensorFlow and includes `venv`. Notice that the `birds` environment uses Python 3.10 explicitly; continue this way for the book's experiments. Now install the remaining toolkits:

```
(birds) > pip3 install tensorflow-macos
(birds) > pip3 install matplotlib
(birds) > pip3 install scikit-learn
(birds) > pip3 install scipy
(birds) > pip3 install pillow
(birds) > deactivate
```

Notice that TensorFlow is macOS-specific. Apple has its own version optimized for Apple hardware. If a GPU is available, `tensorflow-macos` will install for it. If not, a CPU-only version will be installed. If you have GPU hardware, meaning Apple Silicon (M1/M2), you must also install `tensorflow-metal`:

```
(birds) > pip3 install tensorflow-metal
```

Your Mac should now be ready – jump to the next chapter.

3.4   Configuring Windows

Windows users have two options. If you are using later versions of Windows 10 or Windows 11, install the Window Subsystem for Linux (WSL) using the default Ubuntu installation. Then, follow the setup steps above to configure Linux and jump to the next chapter. Otherwise, continue reading to configure older versions of Windows 10.

First, install Python 3.10 from www.python.org. Run the installer and be sure to check "Add Python to PATH". Then, click on "Customize installation" and ensure that `pip` and `venv` are also installed.

When the Python installation completes, open a command prompt and execute:

```
> python -m venv birds
> birds\scripts\activate.bat
(birds) > pip install numpy
(birds) > pip install tensorflow-cpu
(birds) > pip install matplotlib
(birds) > pip install scikit-learn
(birds) > pip install scipy
(birds) > pip install pillow
```

These steps install the necessary libraries. Enter `deactivate` to exit the environment. Installation is now complete, and you're ready for the next chapter.

4. Building a Bird Dataset

Let's build a bird dataset. Of course, we must first have a plan, which involves answering some basic questions: Which bird species? How many images do we need per class and what format should the images be in when we pass them to the model? The answers to these questions imply planning, acquiring and preprocessing, the topics of the chapter's first section.

With the selected images in hand, it's time to build the train and test sets. In this chapter, we won't use a separate validation set but reserve the right to repurpose a portion of the test samples for validation later, should we desire it. The second section accomplishes this step.

The chapter concludes with initial testing of our now fabulous bird dataset and a discussion. Spoiler alert: the results will leave us somewhat dismayed, perhaps even flustered. Indeed, one might even dare to say that our feathers will be a tad ruffled, but have no fear, Chapter 5 gives us hope for a brighter future.

4.1 Planning, Acquiring and Preprocessing

To build the dataset, we must first plan it. Then, once we know what it is we want, we must go through the bother of acquiring the images. After acquiring the images, we'll have the raw data necessary to assemble the dataset. Preprocessing the raw images to transform them into a format appropriate for a deep learning model follows.

4.1.1 Planning and Acquiring

Let's build a dataset consisting of six North American bird species:

(0) American Kestrel
(1) American White Pelican
(2) Belted Kingfisher
(3) Great Blue Heron
(4) Red-tailed Hawk
(5) Snowy Egret

The class label is in parentheses, meaning we label all Great Blue Heron images as class 3 and all Snowy Egrets as belonging to class 5. Examples of each species are in Figure 4.1.

Figure 4.1 Top row (left to right): American Kestrel, American White Pelican, Belted Kingfisher. Bottom row: Great Blue Heron, Red-tailed Hawk, Snowy Egret.

The raw images, too many to make available, come from a collection of photographs I took over a two-year period from three different locations near my home in Thornton, Colorado: Marshall Lake (39.939317, −104.915397), Eastlake (39.927036, −104.951369) and Metzger's Farm (39.916424, −105.029851).

How should we create the dataset from the raw images? Naturally, we need as many images as possible of each of the six species in Figure 4.1. However, care is necessary. We might be tempted to mix the images from all three locations, extract the relevant species, and then partition them into our train and test sets. This is a reasonable first thought, to be sure, but one that is likely to lead to a false impression of the resulting model's generalizability.

Marshall Lake, Eastlake and Metzger's Farm are all within a few miles of each other. Especially close are Marshall Lake and Eastlake, a mere 2 miles or so apart as the crow or egret flies. The probability that the same birds appear in both places is perhaps higher than we'd like. Ideally, we want the data in the training set and test set to be distinct, but from the same overall distribution. In other words, we need the same species, but we don't want the same individual birds between the two, if possible. In this particular case, however, we must live with a less-than-ideal situation as it is likely the same bird appears at times in images from Marshall Lake and Eastlake. We'll tolerate this possibility, and in general, when building image datasets for birds (remember, they can fly!), it might be the case that individuals overlap when partitioning the raw data.

I decided on the following to partition the images into train and test sets: training images from Marshall Lake and test images from Eastlake and Metzger's Farm. This approach will involve overlap but still provide enough train and test data.

Look again at the images in Figure 4.1. The images are reasonable but not sterling regarding pose, coverage, lighting, etc. This is intentional as we cannot know *a priori* the attributes of the images models trained on this dataset will be expected to correctly classify. Indeed, a few of the training images are pretty poor, but that's okay if poor-quality images might be inputs. Again, the training data should reflect what the model will encounter in the wild.

The images in Figure 4.1 are cropped and square. We want square images for this dataset; however, the raw images came out of the camera for whatever zoom factor was used to say nothing about lighting, time of day, time of year, and so on. This implies a level of manual preprocessing to transform the raw images into something from which a dataset might be constructed. It's necessary, even critical, grunt work.

4.1.2 Preprocessing

All the images in this book come from a Canon SX70 HS camera using automatic settings to produce JPEG output. Professional photography was not the goal. The SX70 has a 60× optical zoom with digital zoom out to 230×. Greater zoom implies lower image quality in the optical range due to operator movement, compounded by pixelation effects when switching to digital zoom.

The raw camera images were manually processed to compensate for lighting and color and to crop to remove extraneous portions of the image. Note that this cropping is separate from the cropping necessary to build the dataset.

Images were processed in Gimp (`www.gimp.org`) to adjust color levels, bump the saturation and apply unsharp masking. Will the resulting dataset be sensitive to these adjustments compared to the raw images from the camera? I suspect not because deep networks for computer vision are often quite robust to subtle changes, but I leave answering the question as an exercise for the reader.

Manual processing of the images left them of arbitrary size and aspect ratio. We want square input images for the dataset, so we must crop the images. Not only do we want square images, we want them of a specific size: 256 × 256 pixels ultimately scaled to 64 × 64 and 32 × 32 pixels. Therefore, for each raw image, we must somehow select a square set of pixels centered on the bird (or at least its head region), then rescale the selected pixels to 256 × 256 before writing to disk.

Great! All we must do is select centered square crops and resize them. But how? My answer to this question utilized a custom-built application to perform the following:

1. Load a selected raw image as RGB dropping any alpha channel
2. Wait for the user to click on the image
3. Extract the largest square image with the clicked location as the center
4. Automatically rescale the cropped region to 256 × 256 pixels
5. Store the chip in a user-supplied output directory

Writing such a custom application in Python using wxPython isn't particularly difficult, but it is tedious and requires learning (or remembering) enough of wxPython to accomplish the goal. Fortunately, we no longer need to waste time on such tedious coding tasks: we can ask an AI to do it for us.

I asked OpenAI's GPT-4o to create a custom application to do everything necessary to chip the raw images. GPT-4o's first reply worked but needed tweaking to return to the most recently opened directory when selecting an image. A few additional prompts resolved that issue. I then modified the resulting code to read the output chip size and the name of the output chip directory from the command line. The resulting application is in *chipper.py*. To run it, first install wxPython:

```
> pip3 install wxpython
```

then run *chipper.py* without arguments:

```
> python3 chipper.py

chipper <size> <outdir>

    <size> - chip output size (always square)
    <outdir> - output chip directory
```

Executing the code with

```
> python3 chipper.py 256 chips
```

produces a blank window with a *File* menu from which the user selects the first image to chip.

The selected image is loaded, dropping any alpha channel, and displayed in the window. The application then waits for the user to click on the image before extracting the largest square region centered on the location of the mouse click, rescaling the region to 256×256 pixels and storing the rescaled chip in the output chip directory. Selecting a new source image returns to the previous directory. Clicking the image a second time selects a new chip overwriting any previously selected chip. Both behaviors enable a reasonably rapid workflow where raw images are quickly chipped and saved before choosing the next.

I ran *chipper.py* twice, first to select previously segregated training set images from Marshall Lake for each of the six species. The results were chipped to the *train* directory. The second run did the same for images from Eastlake and Metzger's Farm to form the test set in the *test* directory. This exercise, which was tedious but not overly taxing, produced a collection of 413 images in the training set and 396 in the test set, as indicated in Table 4.1.

The table lists the number of train and test images for each species. A cursory glance tells us that the datasets are not balanced, meaning there is considerable variation in the number of images by class. For example, the training set contains over 100 Belted Kingfisher images but only a meager 37 Red-tailed Hawk images. The test set is similarly imbalanced but in the opposite direction: 133 Red-tails to only 23 Kingfishers. We'll return to this class disparity in Chapter 5.

Let's pause and consider how much effort has gone into our dataset so far. First, someone (meaning me) had to spend countless hours walking, camera in tow, for two years to acquire the raw images. We might call these the "proto-raw" images as each image selected for the dataset was itself manually edited, as indicated earlier. Add dozens of additional hours for that. Finally, transforming the raw images into two collections of

Table 4.1 The *bird6* train and test counts by class

Species	Train	Test
(0) American Kestrel	74	85
(1) American White Pelican	42	29
(2) Belted Kingfisher	103	23
(3) Great Blue Heron	74	80
(4) Red-tailed Hawk	37	133
(5) Snowy Egret	83	46

256×256 pixel RGB chips required a creating a custom application followed by using said application; increment the tally by an additional couple of hours.

I'm not fishing for sympathy – indeed, birding while walking is often the highlight of my day. I make mention only to reinforce my claim that building datasets takes effort, and is often the majority of what goes into designing a successful model. We expect models to learn to generalize from data, so this observation shouldn't be surprising – but it's worth remembering.

We have two directories, *train* and *test*, each with approximately 400 bird images. We're now ready for the next step: bundling the images and additional rescaling to produce datasets small enough for the models we seek to train.

4.2 Building Train and Test Sets

The previous section separated train and test for us. What remains is to bundle the images into individual NumPy files giving us access to the images as an array in memory. For small datasets, this is typically the best approach. Large datasets that do not fit into memory require more scaffolding.

The necessary code is in *build_bird6.py*. Please review it before continuing. Listing 4.1 contains the main code with imports excluded.

```
np.random.seed(19937)

train_names=np.array(["train/"+i for i in os.listdir("train")])
idx = np.argsort(np.random.random(len(train_names)))
train_names = train_names[idx]

test_names = np.array(["test/"+i for i in os.listdir("test")])
idx = np.argsort(np.random.random(len(test_names)))
test_names = test_names[idx]

train_labels = np.array([ExtractLabel(i) for i in train_names])
test_labels = np.array([ExtractLabel(i) for i in test_names])

x64,x32,g64,g32 = ExtractImages(train_names)
np.save("../data/bird6_64_xtrain.npy", x64)
np.save("../data/bird6_32_xtrain.npy", x32)
np.save("../data/bird6_gray_64_xtrain.npy", g64)
np.save("../data/bird6_gray_32_xtrain.npy", g32)
np.save("../data/bird6_ytrain.npy", train_labels)

x64,x32,g64,g32 = ExtractImages(test_names)
np.save("../data/bird6_64_xtest.npy", x64)
np.save("../data/bird6_32_xtest.npy", x32)
np.save("../data/bird6_gray_64_xtest.npy", g64)
np.save("../data/bird6_gray_32_xtest.npy", g32)
np.save("../data/bird6_ytest.npy", test_labels)
```

Listing 4.1 Building the *bird6* dataset

The code is split into six paragraphs. The first sets NumPy's pseudorandom number seed to order the data consistently, run to run (this step is not strictly necessary).

The following two paragraphs define `train_names` and `test_names` as lists of the train and test filenames, respectively. These are the names *chipper.py* used when writing to disk. Notice that `idx` is used to scramble the order of the names to avoid the sorted order returned by the call to `listdir`.

The next paragraph calls `ExtractLabel` repeatedly, passing it each chip filename, to build a list of class labels. The function `ExtractLabel` matches the image name to species names to determine the class; see Listing 4.2.

```
def ExtractLabel(name):
    if      ("kestrel" in name):    ans=0
    elif    ("pelican" in name):    ans=1
    elif    ("kingf" in name):      ans=2
    elif    ("heron" in name):      ans=3
    elif    ("red-tail" in name):   ans=4
    elif    ("snowy" in name):      ans=5
    else:
        raise ValueError("Unknown class: %s" % name)
    return ans
```

Listing 4.2 Mapping names to labels

The final two code paragraphs in Listing 4.1 call `ExtractImages` to build RGB and grayscale versions of the train and test datasets, one each for images rescaled to 64×64 and 32×32. The original 256×256 images are too large for the simple models we will explore, but the larger images are likely what should be used in a research setting (with available GPUs). Listing 4.3 contains `ExtractImages`.

```
def ExtractImages(names):
    N = len(names)
    x64 = np.zeros((N,64,64,3), dtype="uint8")
    x32 = np.zeros((N,32,32,3), dtype="uint8")
    g64 = np.zeros((N,64,64), dtype="uint8")
    g32 = np.zeros((N,32,32), dtype="uint8")
    for i in range(N):
        img = Image.open(names[i]).convert("RGB")
        i64 = img.resize((64,64), resample=Image.BILINEAR)
        i32 = img.resize((32,32), resample=Image.BILINEAR)
        x64[i,...] = np.array(i64)
        x32[i,...] = np.array(i32)
        g64[i,...] = np.array(i64.convert("L"))
        g32[i,...] = np.array(i32.convert("L"))
    return x64,x32,g64,g32
```

Listing 4.3 Reading and cropping images

Extracting images involves a loop over the given filenames to read the image from disk, ensuring that it is RGB only (no alpha channel), then rescaling to 64 × 64 and 32 × 32 pixels. The respective images are stored in x64 and x32 before placing grayscale versions (mode "L") in g64 and g32, all of which are then returned.

Run the code to build the dataset:

```
> python3 build_bird6.py
```

The output files are in the *data* directory. The different image formats are in the *xtrain* and *xtest* NumPy files with the corresponding labels in the *ytrain* and *ytest* files.

4.3 Initial Testing

Let's train versions of our basic models using the *bird6* dataset. The code we need is in *train_bird6.py*. It serves as a template for future experiments. Run the code without arguments to learn what's expected:

```
train_bird6 <mname> rgb|gray 64|32 <minibatch> <epochs> <outdir>

    <mname>      - model: lenet5|vgg8|resnet18
    rgb|gray     - RGB or grayscale
    64|32        - 64x64 or 32x32 images
    <minibatch>  - minibatch size (e.g. 64)
    <epochs>     - number of training epochs (e.g. 16)
    <outdir>     - output directory name
```

As an example, let's train a simple LeNet-5 model using 32 × 32 grayscale images for 64 epochs with a minibatch of 16. As we'll see, the code reserves 15 percent of the test data for validation. We'll dump the output to a temporary directory:

```
python3 train_bird6.py lenet5 gray 32 16 64 tmp
```

Keras gives us status while the model trains, epoch by epoch, indicating the training loss, training accuracy, validation loss and validation accuracy. For example, consider

```
Epoch 64/64
26/26 loss: 0.0227 - accuracy: 0.9927 -
        val_loss: 3.0786 - val_accuracy: 0.4407
```

where I've removed some status information to focus on the losses and accuracies. The example presents the results for the final training epoch. The 26/26 indicates there are 26 minibatches per epoch with a final training set accuracy of 99.27 percent and a corresponding validation set accuracy of 44.07 percent. As the model learns, the training set accuracy will approach 100 percent while the validation accuracy will often plateau and, if overfitting, begin to increase with epoch.

The test set confusion matrix and overall accuracy is given when training is complete. In this case it was:

```
[[50  2  4  3  3  8]
 [ 0  8  1  1  1 12]
 [ 2  2 14  2  0  0]
 [11 10  7 18  9 15]
 [39  4 17  9 28 18]
 [ 1  8  4  4  1 21]]
Test set accuracy: 0.4125, MCC: 0.3083
```

There are six classes in the dataset. The confusion matrix, therefore, has six rows and six columns, with each row the true class label, $[0, 5]$, and each column the model's assigned label. The first row indicates the model's performance on class 0, American Kestrels:

```
[50  2  4  3  3  8]
```

The test set, excluding the portion held back for validation, contains 70 kestrel images (row sum). The LeNet-5 model correctly identified 50 of them as American Kestrels for an accuracy of $50/70 \approx 0.714$ or 71.4 percent. That's not a fantastic result, nor is it pathetic, but recall that the images are grayscale and only 32×32 pixels.

Now consider row 4 corresponding to class 4, Red-tailed Hawks:

```
[39  4 17  9 28 18]
```

Here the model was less successful. The test set contained 115 red-tail images, of which the model correctly identified $28/115 \approx 0.243$ or 24.3 percent. The remaining red-tail images were assigned to other categories with the majority, 33.9 percent, declared to be kestrels instead. The model was at least mainly unwilling to call red-tails pelicans (4/115) or herons (9/115). That fact combined with believing red-tails to be kestrels indicates to me that the model did learn something useful, that is, it wasn't guessing. Kestrels are, after all, birds of prey, like red-tails. Still, it's a poor showing overall with an accuracy of 41 percent across all classes.

The *tmp* directory contains several files:

```
accuracy_mcc.txt
command_line.txt
confusion_matrix.npy
error_plot.png
model.keras
```

The first (*accuracy_mcc.txt*) contains the test set accuracy line in the output. The full command line is in *command_line.txt*, and the confusion matrix as a NumPy array is in *confusion_matrix.npy*. The trained model is in *model.keras*. A plot of the train and validation error (1 − accuracy) is also present and shown in Figure 4.2.

The plot tells us that the training error decreases with epoch, as it does because we are using gradient descent, so we will eventually reach a low training error. What is more revealing is the validation error line. It is noisy, reflecting the small validation set

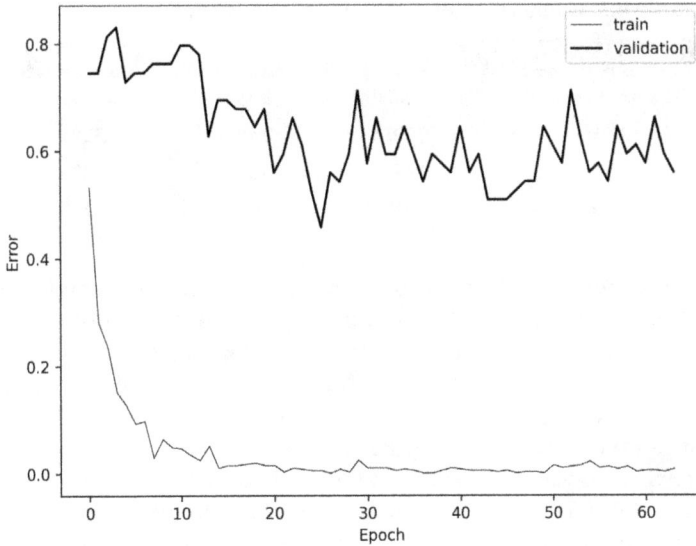

Figure 4.2 The *bird6* LeNet-5 error plot

(15 percent of the test data), and it decreases initially but then levels off around epoch 40 or so, indicating that the model has likely learned all that it will learn.

Let's repeat the exercise using 64 × 64 grayscale images:

```
python3 train_bird6.py lenet5 gray 64 16 64 tmp
```

Intuitively, we expect the larger images to produce a better model; after all, the images are four times the size and therefore contain significantly more detail. However, my run produced the following output:

```
[[53    2    8    5    1    1]
 [ 0  10    4    6    0    3]
 [ 7    1  10    2    0    0]
 [16    3  13  31    4    3]
 [46    7  24  24  12    2]
 [ 1    8    6  13    0  11]]
Test set accuracy: 0.3769, MCC: 0.2731
```

which is significantly less accurate overall than the 32 × 32 case. Comparing the diagonals between the 64 × 64 and 32 × 32 confusion matrices offers a clue:

```
64×64:  53   10   10   31   12   11
32×32:  50    8   14   18   28   21
```

The diagonal represents correct classifications. The 64 × 64 model improved over the 32 × 32 model for kestrels (0), pelicans (1) and herons (3), but was considerably worse at classifying kingfishers (2), red-tails (4) and egrets (5). The 64 × 64 model was especially bad at identifying red-tails with most of them falling into the kestrel camp.

Table 4.2 Overall test set accuracies by model and data type

	LeNet-5	VGG8	ResNet-18
32 × 32 grayscale	0.3680	0.3828	0.2493
32 × 32 RGB	0.3976	0.4807	0.3294
64 × 64 grayscale	0.4451	0.5015	0.3234
64 × 64 RGB	0.4481	0.3976	0.3709

The 64 × 64 confusion matrix and the 32 × 32 result represent single training runs. We know models are randomly initialized, and repeated training produces models with different performances. This is especially true for small datasets. However, even with this caveat, it appears that the LeNet-5 model, for grayscale images, is better suited to the smaller 32 × 32 images. This is perhaps due to the size of the model and the range of functions it can represent. The smaller images contain simpler representations of the differences between classes and the LeNet-5 model is perhaps better able to learn these representations while it lacks the capacity to do as well with the larger 64 × 64 models. To tease out the validity or falsehood of these statements requires training many models to generate relevant statistics followed by hypothesis tests.

The script file *train_bird6_models* trains all combinations of RGB vs grayscale, 64 × 64 vs 32 × 32 and model type (LeNet-5, VGG8, ResNet-18). Execute it with:

```
> sh train_bird6_models
```

My run took about 80 minutes to complete. The resulting model accuracies are in Table 4.2. Remember that each represents a single training session, so running a second time will produce different results, but general trends should persist.

Table 4.2 is difficult to interpret rigorously. Overall, it appears that the LeNet-5 and VGG8 models are generally better performing than ResNet-18, likely due to the size of the training set (ResNets favor larger training sets), but it is not possible to choose between LeNet-5 and VGG8, nor between RGB and grayscale, without many additional training runs.

4.4   Reviewing the Code

Let's conclude the chapter by walking through *train_bird6.py* to understand each part. The code serves as a guide to future experiments. The code begins by importing necessary libraries, including the model-specific ones from Chapter 2:

```
from lenet5    import LeNet5, ConfusionMatrix
from vgg8      import VGG8
from resnet18  import ResNet18
```

Next, the command line is parsed:

```
mname = sys.argv[1].lower()
itype = sys.argv[2].lower()
isize = int(sys.argv[3])
batch_size = int(sys.argv[4])
epochs = int(sys.argv[5])
outdir = sys.argv[6]
```

followed by loading the proper train and test data according to the image size and type. For example,

```
if (isize == 64):
    if (itype == "rgb"):
        x_train = np.load("../data/bird6_64_xtrain.npy")
        x_test = np.load("../data/bird6_64_xtest.npy")
        input_shape = (64,64,3)
ytrain = np.load("../data/bird6_ytrain.npy")
ytest = np.load("../data/bird6_ytest.npy")
num_classes = 6
```

Notice the definitions of `input_shape` and `num_classes`. The train and test arrays are of shape:

```
>>> x_train.shape
(413, 64, 64, 3)
>>> x_test.shape
(396, 64, 64, 3)
```

to reflect the size of each dataset (413 vs 396 images), the size of each image, and the number of image channels, 3 for RGB.

The images are still in their native [0, 255] byte format. We want to scale them to [0, 1] by dividing by 255. Also, the labels are integers, [0, 5], but Keras needs them to be one-hot vectors as the model output is a softmax vector:

```
x_train = x_train.astype('float32') / 255
x_test = x_test.astype('float32') / 255

y_train = keras.utils.to_categorical(ytrain, num_classes)
y_test = keras.utils.to_categorical(ytest, num_classes)
```

Next, 15 percent of the training data is set aside to serve as validation data during training, even though, in this case, we are not yet using the validation data to decide when to stop training:

```
n = int(0.15*len(y_test))
x_test, x_val = x_test[n:], x_test[:n]
y_test, ytest, y_val = y_test[n:], ytest[n:], y_test[:n]
```

Preprocessing is now complete, implying we're ready to define the model:

```
if (mname == "lenet5"):
    model = LeNet5(input_shape, num_classes)
elif (mname == "vgg8"):
    model = VGG8(input_shape, num_classes)
else:
    model = ResNet18(input_shape, num_classes)
```

using `input_shape` and `num_classes`, then compile and train:

```
model.compile(loss=keras.losses.categorical_crossentropy,
              optimizer=keras.optimizers.Adam(),
              metrics=['accuracy'])

history = model.fit(x_train, y_train,
              batch_size=batch_size,
              epochs=epochs,
              verbose=1,
              validation_data=(x_val, y_val))
```

Training uses the advanced Adam optimizer and tracks accuracy, epoch by epoch. The validation data is supplied (`x_val`) and `verbose=1` ensures feedback during training.

Finally, the trained model is tested against the portion of the test data not used for validation:

```
pred = model.predict(x_test, verbose=0)
plabel = np.argmax(pred, axis=1)
cm, acc = ConfusionMatrix(plabel, ytest,
              num_classes=num_classes)
mcc = matthews_corrcoef(ytest, plabel)
txt = 'Test set accuracy: %0.4f, MCC: %0.4f' % (acc,mcc)
```

followed by displaying the confusion matrix, the overall accuracy with Matthews correlation coefficient (MCC) and dumping results to the output directory (not shown).

We must understand the difference between what is in `pred` and what is in `plabel`. The former is the model's softmax output, a six-element vector in [0, 1] for each image, where each element is the model's confidence of membership in that class. It is not a class label.

Transforming the softmax predictions into class labels requires learning the index of the largest softmax value. For instance, a possible softmax output for a given input might be

$$0.0741 \quad 0.4074 \quad 0.1111 \quad 0.2222 \quad 0.1481 \quad 0.0370$$

which sums to 1.0 as probabilities must (to within rounding). The biggest value is 0.4074 for class 1 (index 1) implying the model is most confident that the input image is of a pelican. Calling NumPy's `argmax` function with `axis=1` returns the index of the largest element in each row of the `pred` matrix, precisely what is needed to map softmax vectors to class labels in `plabel`.

A maximum softmax value of 0.4074 isn't a ringing endorsement of class membership. High-quality models often produce softmax outputs of 0.9 or above to indicate a strong belief in membership. We are colloquially calling softmax vectors "probabilities" because they sum to 1.0 across the classes, but it's critical to remember that unless the model is externally calibrated, itself an active research area, the softmax values are more correctly interpreted as likelihoods, meaning a softmax value of 0.9 should not be immediately interpreted as a 90 percent probability of membership in the class, only that the model strongly "believes" that that is the likely class. In Chapter 5, we'll explore how we might use the softmax vector and selected threshold values to decide whether to assign a class label or not.

4.5 Discussion

This chapter saw us construct a small, six-class bird dataset using a pool of existing images. Most of the effort, beyond acquiring the images in the first place, took place in moving from raw images to cropped images, the manually intensive portion of the task. A custom application, AI generated, proved essential.

Processing the cropped images required a few dozen lines of Python code. Training models involved a similar amount of code to load the requested datasets (train and test), partition to create a tiny validation set, and define the desired model after scaling the input data to [0, 1].

Training at this point is nothing more than two method calls on the `model` object: one to `compile` to inform Keras (and via it TensorFlow) about the training process, and another to `fit` to train for the requested number of epochs.

Inference with a trained model means a call to `predict` passing in a set of unknown bird images, all assumed to be within the six classes the model understands (more on that in Chapter 5). Test data includes the labels, thereby allowing verification of the model's predictions.

The next chapter continues exploring the *bird6* dataset to uncover its weaknesses before attempting to address them in a meaningful way.

5. Exploring the *Bird6* Dataset

This chapter explores the *bird6* dataset from Chapter 4. Our goal is to gain experience with models, especially models trained from scratch using small datasets. Many of deep learning's successes in recent years come from *foundation models*, large (deep) models trained on extensive datasets. In practice, we usually find ourselves on the opposite end of the deep learning pool with small datasets and small models. The experiments of this chapter increase your experience in this area, reveal new approaches to handling small models, and prepare us for the next chapter which introduces pretrained models, transfer learning and fine-tuning.

The first section briefly explores the effect of hyperparameters on the training process. We could easily spend the entire chapter on this topic by diving into the possible parameters and studying their effects, but we'll restrict ourselves here to two key hyperparameters: minibatch size and number of epochs.

The following section introduces data augmentation in an effort to improve the models trained on the *bird6* dataset. Data augmentation is a powerful technique and should feature prominently in your machine-learning toolkit.

Models produce softmax vectors as output from which a class label is selected. It's therefore worth our time to explore the raw model output to uncover situations where we might decide not to select a class label but to instead refuse to make a decision, thereby cueing an outside observer to pay closer attention to a particular set of inference-time inputs. The same logic applies to cases where the model is subjected to inputs outside the set on which it was trained, that is *out-of-distribution (OOD)* inputs.

The final section before the closing discussion explores the effect of *ensembles*, collections of separately trained models that jointly assign an inference-time sample to a particular class.

5.1   Exploring Hyperparameters

Training involves updating the initial set of model weights and biases via gradient descent. Typically, training takes a gradient descent step after processing each minibatch. The minibatch samples are passed through the model to produce a set of predictions using the current set of model weights and biases (the *forward pass*). The loss function then determines the error between the predictions and the known labels. Backpropagation is used to determine the partial derivatives reflecting each weight and biases' influence on the minibatch error (the *backward pass*). Gradient descent takes these partial derivatives, multiplied by a learning rate to adjust the current model parameters – in other words, to take a step through the loss landscape. Optimizers like Adam adjust learning rates on a per-parameter basis.

This section explores the relationship between minibatch size, training set size, epochs and gradient descent steps to help us understand how they work together to train a model.

Assume there are N samples in a training set and the minibatch size is m. For simplicity, assume m evenly divides N so that the number of minibatches in the dataset is N/m with each training sample used once per epoch.

The network's error on a minibatch is used to update the model's weights and biases, therefore, each minibatch maps to a gradient descent step implying N/M such steps in one epoch and

$$s = e\left(\frac{N}{m}\right)$$

training steps, s, over a given number of epochs, e.

Keras lets us specify the minibatch size and number of epochs. As an experiment, let's repeatedly train a LeNet-5 model for a fixed number of gradient descent steps, s, while varying the minibatch size, m. If we change m, we must also change e so that $e(N/m)$ remains constant.

The file *lenet5_tests.py* runs the experiment using calls to *train_bird6.py* varying m and e as needed (N, the training set size, is fixed). For each m and e, 12 models are trained (LeNet-5, grayscale images, and 32 × 32 or 64 × 64 images) tracking the overall accuracy of each. Multiple runs give us a mean accuracy for each set of minibatch and epochs, thereby controlling for randomness introduced by network initialization.

Minibatch size varies as

$$m = \{1, 2, 4, 8, 16, 32, 64, 128\}$$

implying

$$e = \{2, 4, 8, 16, 32, 64, 128, 256\}$$

so that $e/m = 2$, always. Fixing the ratio tells us that $s = 2N$ gradient descent steps during training. For *bird6*, $N = 413$, implying training will always take $s = 2N = 2(413) = 826$ gradient descent steps.

If we were to change m without changing e, the number of gradient descent steps would decrease as m increases, thereby confounding what we want to explore: the effect of minibatch size, all other things remaining equal.

I'll let you review the code in *lenet5_tests.py*. When run, it produces a summary indicating the relationship between m and overall accuracy for the different image sizes. My run produced:

```
( 1,  2)    0.160354 +/- 0.019021   (0.145202)
( 2,  4)    0.233165 +/- 0.013067   (0.236111)
( 4,  8)    0.378998 +/- 0.013572   (0.382576)
( 8,16)     0.390783 +/- 0.013606   (0.405303)
(16,32)     0.395623 +/- 0.008865   (0.386364)
(32,64)     0.397727 +/- 0.007724   (0.402778)
(64,128)    0.403409 +/- 0.007123   (0.402778)
(128,256)   0.375421 +/- 0.015270   (0.386364)
```

for grayscale 32×32 images and

```
( 1,  2)    0.178662 +/- 0.018684   (0.170455)
( 2,  4)    0.194655 +/- 0.013795   (0.200758)
( 4,  8)    0.377525 +/- 0.009364   (0.385101)
( 8, 16)    0.375000 +/- 0.010358   (0.375000)
(16, 32)    0.389941 +/- 0.007186   (0.383838)
(32, 64)    0.403199 +/- 0.006388   (0.401515)
(64,128)    0.400673 +/- 0.004593   (0.398990)
(128,256)   0.401305 +/- 0.008470   (0.397727)
```

for 64×64 images. The mean accuracy (± SE) is given for each (m,e) combination with the median in parentheses. Let's focus for a moment on the 32×32 results.

A minibatch of one ($m = 1$), often called the *online learning* case, estimates the loss function gradient from a single training sample. As such, we shouldn't be surprised that the mean performance of models trained in that way is low, with about 18 percent accuracy on the test set. Recall that six classes imply a random guess accuracy of 16.7 percent, so the $m = 1$ case is barely learning.

The minibatch should be representative of the training set as a whole, as a political poll is (supposed to be) representative of the population as a whole. A single training sample cannot meet this requirement, so the models learn poorly.

Overall accuracy increases as m increases reaching a peak of 40.3 percent when $m = 64$ and $e = 128$. Therefore, we have evidence that the best minibatch size to use when training a LeNet-5 model on the 32×32 grayscale version of *bird6* is 64 samples. Notice how increasing the minibatch further results in reduced accuracy (only 37.5 percent). This effect is commonly observed and supports the claim that stochastic gradient descent based on a noisy but reasonable estimate of the true gradient (which cannot actually be known) often leads to the best models. The estimate of the true generalization loss when $m = 128$ is likely closer to the unknowable true loss, but such a level of accuracy allows training to fall into local minima or saddle points (areas of a function that are flat, but not true minima). A similar effect is observed for the 64×64 images with a peak accuracy when $m = 128$.

It's natural to wonder if $m = 64$ and $m = 128$ are genuinely the "best" minibatch sizes. Run *lenet5_analysis.py* to compare the results for these minibatches to the others using t-tests and nonparametric Mann-Whitney U tests. If the resulting p-values are significantly below 0.05 (by an order or magnitude), we might come to believe that they are. However, the hypothesis tests show that only $m = 1$ and $m = 2$ produce truly inferior models, as is plainly evident from the resulting accuracies on the test data.

Let's combine all this to ensure we capture what is happening. We tested multiple minibatch sizes while altering the number of training epochs so that the total number of gradient descent steps (weight and biases updates) remains constant. Various models were trained for each minibatch size ($n = 12$) so that we have some confidence in the resulting mean accuracies. Observing these accuracies told us there is a "sweet spot" minibatch size in the 32 to 128 range. Hypothesis tests indicate that there is little meaningful difference over this range, which is commonly observed with other datasets. Therefore, as a rule of thumb, we should consider minibatch sizes of 32 to 64 and refine the number only when attempting to fine-tune the training process.

This experiment is a sobering assessment of the *bird6* dataset. Frankly, it doesn't lead to well-performing models. An overall accuracy of 40-some percent isn't likely to be valuable enough in practice. The dataset is simply too small, especially for classes like red-tails with scarcely three dozen training examples. Fortunately, we have some tricks up our sleeve that might help matters.

The best thing to do is get more training data, at least an order of magnitude (or two) more. However, reality says that we're stuck with what we have. However, what if we could *pretend* to have more data than we actually do? That's the topic of the next section.

5.2   Data Augmentation

We know that more training data improves model performance, so, let's invent some. Data augmentation does just that, but in a principled way that reflects reality, not the usual Silicon Valley way of "fake it 'til you make it."

Data augmentation takes existing training samples, here images, and produces new versions of them that are different enough from the original to be "perceived" by the model as new, yet similar enough to be a plausible instance of the class. For images, data augmentation typically involves image transformations like rotations, flips, horizontal or vertical shifts or scaling to zoom in or out. We learned about such augmentations in Chapter 2. Let's put these augmentations to work to expand the training set and, hopefully, improve the quality of our models.

The previous section trained images with *train_bird6.py*, code we developed in Chapter 4. The code in *train_bird6_augment.py* is fundamentally the same but inserts an augmentation step acting before training begins. It is typical to augment the training set, though it is possible to augment the test set as well, but we won't do that here because we already have nearly 400 test samples. The relevant new code in *train_bird6_augment.py* is found in the functions Augment and AugmentDataset; see Listing 5.1.

```
def Augment(im):
    img = Image.fromarray(im)
    if (np.random.random() < 0.5):
        img = img.transpose(Image.FLIP_LEFT_RIGHT)
    if (np.random.random() < 0.5):
        r = 3*np.random.random()-3
        img = img.rotate(r, resample=Image.BILINEAR)
    if (np.random.random() < 0.5):
        i = np.array(img)
        n = int(0.1*i.shape[0])
        i = np.roll(i, np.random.randint(-n,n+1), axis=1)
        i = np.roll(i, np.random.randint(-n,n+1), axis=0)
        img = Image.fromarray(i)
    return np.array(img)

def AugmentDataset(x,y, factor=10):
    if (x.ndim == 3):
        n, height, width = x.shape
        newx = np.zeros((n*factor, height, width),
                    dtype="uint8")
```

```
        else:
            n, height, width, channels = x.shape
            newx = np.zeros((n*factor, height, width, channels),
                        dtype="uint8")
        newy = np.zeros(n*factor, dtype="uint8")
        k=0
        for i in range(n):
            im = Image.fromarray(x[i,:])
            newx[k,...] = np.array(im)
            newy[k] = y[i]
            k += 1
            for j in range(factor-1):
                newx[k,...] = Augment(x[i,:])
                newy[k] = y[i]
                k += 1
        idx = np.argsort(np.random.random(newx.shape[0]))
        return newx[idx], newy[idx]
```

Listing 5.1 Augmenting training images

The function `AugmentDataset` loops over the samples calling `Augment` on each image `factor` times. Notice that the original, unaugmented image is kept, followed by additional augmented versions, and each augmented image receives the same label as the original, as it must be. After scrambling, the augmented dataset is returned to ensure that minibatches are consistently diverse.

All the fun happens in the `Augment` function. The input image is randomly flipped left to right, rotated $[-3, +3]$ degrees, or shifted up to 10 percent of its height and width, all with a probability of 0.5. In other words, the set of applied transforms varies on each call to `Augment`. Notice that we are not scaling the images or altering the gamma or color content, though if desired, `Augment` is the place to add the necessary code.

Augmentation happens after loading the desired training set and before scaling and creating the one-hot label vectors:

```
x_train, ytrain = AugmentDataset(x_train, ytrain, factor)

x_train = x_train.astype('float32') / 255
x_test = x_test.astype('float32') / 255

y_train = keras.utils.to_categorical(ytrain, num_classes)
y_test = keras.utils.to_categorical(ytest, num_classes)
```

Augmentation is automatic, requiring only the augmentation factor as a new command line argument. For instance,

```
python3 train_bird6_augment.py lenet5 rgb 32 16 32 20 tmp
```

will train a LeNet-5 model using 32×32 RGB images, a minibatch of 16, 32 epochs and an augmentation factor of 20 dumping output in the *tmp* directory. An augmentation factor of

20 increases the training set from 413 samples to 8260, the original 413 plus 19 augmented versions of each making the new dataset 20× larger. Notice that augmentation is applied equally to each class, thereby preserving the relative frequency with which each appears in the training data. It is possible to augment only specific classes, like those that are under-represented, but we will not explore that option here.

The script *train_bird6_augment_models* trains all combinations of models, image types and image sizes using 20× augmentation. This matches the output of the script *train_bird6_ models* from Chapter 4 and gives us a direct comparison between the two, provided we remember the caveat that single training runs might be quite different due to initialization randomness. Executing the script took about 26 hours on my test machine, so patience (or a GPU) is helpful.

The file *compare_results.py* compares results from Chapter 4 (unaugmented) to the augmented models presenting the change in overall accuracy followed by side-by-side confusion matrices. First, changes in overall accuracy:

```
rgb:32:    lenet5: ACC: 0.39760  =>  0.54300
rgb:32:      vgg8: ACC: 0.48070  =>  0.52230
rgb:32: resnet18: ACC: 0.32940  =>  0.52820
rgb:64:    lenet5: ACC: 0.44810  =>  0.55190
rgb:64:      vgg8: ACC: 0.39760  =>  0.56680
rgb:64: resnet18: ACC: 0.37090  =>  0.56970
gray:32:   lenet5: ACC: 0.36800  =>  0.50150
gray:32:     vgg8: ACC: 0.38280  =>  0.52520
gray:32: resnet18: ACC: 0.24930  =>  0.46590
gray:64:   lenet5: ACC: 0.44510  =>  0.48370
gray:64:     vgg8: ACC: 0.50150  =>  0.50740
gray:64: resnet18: ACC: 0.32340  =>  0.54010
```

The arrow (=>) shows before and after for accuracy and MCC (not shown). In every case, regardless of image type, size and model type, augmentation improved the model reaching a high of nearly 57 percent for the largest color images (RGB, 64 × 64), a full 20 percentage point increase over the unaugmented model. Notice, also, that the improvement was most pronounced for ResNet-18 models, validating the statement that ResNet models work best with bigger datasets.

Augmentation improved the model overall, but how did it affect individual class accuracies, especially the difficult Red-tailed Hawk class with the tiniest number of training samples? Let's compare the confusion matrices for the ResNet-18 model trained on 64 × 64 RGB images before and after augmentation:

$$
\begin{bmatrix}
25 & 1 & 20 & 6 & 18 & 0 \\
0 & 16 & 2 & 1 & 0 & 4 \\
1 & 0 & 17 & 1 & 1 & 0 \\
11 & 5 & 26 & 15 & 11 & 2 \\
27 & 4 & 40 & 12 & 32 & 0 \\
1 & 16 & 1 & 1 & 0 & 20
\end{bmatrix}
\quad
\begin{bmatrix}
47 & 1 & 3 & 5 & 13 & 1 \\
0 & 9 & 2 & 8 & 0 & 4 \\
0 & 0 & 17 & 2 & 1 & 0 \\
8 & 0 & 6 & 36 & 19 & 1 \\
29 & 0 & 5 & 18 & 60 & 3 \\
1 & 3 & 0 & 12 & 0 & 23
\end{bmatrix}
$$

The unaugmented confusion matrix is on the left and the augmented on the right. Recall that the test set was unaugmented allowing direct comparison between the models. The shaded row shows us how the respective models assigned class labels for the 115 red-tail inputs in the test set.

The unaugmented model (left) correctly labeled 32 of the hawks, called 27 of them kestrels and a whopping 40 of them kingfishers. However, the model trained on augmented data correctly labeled 60 red-tails, while still dumping 29 in the kestrel category. However, the obsession with kingfishers is now gone with only 5 hawks so called. Clearly, data augmentation was helpful in this case.

Overall, performance in every class either improved or remained the same except for class 1 (pelicans), where performance fell with many pelicans identified as Great Blue Herons instead. The unaugmented training set contains 42 pelican examples, only 5 more than red-tails, potentially explaining the noisy performance even when augmented. As always, more data implies better regularization of the model and also better results at inference time.

The previous results with an augmentation factor of 20× demonstrate a clear improvement in model performance. It's natural to expect further augmentation to result in even better models. The file *augment_models.py* tests this hypothesis for augmentation factors up to 150× (64 RGB images, LeNet-5 model). Please review the code to understand how it works.

The code takes some time to run, about 13 hours on my CPU-only test machine. When complete, it presents the mean±SE test set accuracy and MCC ($n = 6$) along with the corresponding median in parentheses. The accuracy results for my run were as follows:

```
(   5)   0.480712 +/- 0.011076   (0.486647)
(  10)   0.502473 +/- 0.013591   (0.504451)
(  15)   0.492087 +/- 0.012506   (0.483680)
(  20)   0.536597 +/- 0.007791   (0.532641)
(  25)   0.525717 +/- 0.016932   (0.529674)
(  50)   0.532146 +/- 0.020682   (0.522255)
(  75)   0.515826 +/- 0.004810   (0.516320)
( 100)   0.549456 +/- 0.017156   (0.543027)
( 150)   0.552918 +/- 0.006665   (0.559347)
```

Accuracy improves, generally, but not consistently. Randomness plays multiple roles here, affecting model initialization and augmentation, which is performed on the fly. Still, it's fair to say that further augmentation helps up to a point. I chose the LeNet-5 model because it is small and trains (relatively) quickly, but because it is small, there is likely a point where training set diversity, only in matters of position because of the selected augmentations, can no longer be utilized by the model.

Augmentation helps, but the examples in this section still assign every input to a known class. It would be nice to have the option of refusing to assign a label to likely incorrect outputs. To do that requires exploring the concept of a *decision threshold*.

5.3   Decision Thresholds

Standard practice dictates assigning the label of the biggest softmax vector value when classifying an unknown input to a model. Most of the time, this is exactly what we want, doubly so if the model is known to be highly accurate. However, there are times when it's better for the model to refuse to assign a label if there is reasonable doubt about accuracy. Our *bird6* model, even when trained on an augmented dataset, is such an example. We'd like to have some notion of the level of faith we should place in the model's declarations. In other words, we'd like to have one or more *decision thresholds* above which we label the input and below which we refuse. This section explores selecting decision thresholds and how we might use them to detect OOD inputs.

5.3.1 Selecting Decision Thresholds

Output softmax vectors give us what we need to select a decision threshold. The simplest approach considers only the maximum softmax value reasoning that there is a value above which the model is more likely to be correct than not. It's well-known that neural networks are often overly optimistic in their likelihood values, which is why it's a touch sloppy to refer to a softmax vector as a set of probabilities, but as long as that class-level optimism is consistent, we can use it to make a decision based on a cutoff value.

Consider the file *thresholds.py*, which accepts an image type and size followed by the name of a model trained on *bird6*. The code loads the proper test data and then splits it in half. The first half is used to find a threshold value, and the second half is used to test the threshold value. Listing 5.2 picks up the story after loading the test images and labels as xtest and ytest, respectively.

```
model = load_model(mname)

n = len(ytest)//2
ytest0, ytest1 = ytest[:n], ytest[n:]
xtest0, xtest1 = xtest[:n], xtest[n:]

pred = model.predict(xtest0, verbose=0)
plabel = np.argmax(pred, axis=1)

pc = pred[np.where(plabel==ytest0)].max(axis=1)
pw = pred[np.where(plabel!=ytest0)].max(axis=1)

threshold = Search(pc,pw)

pred1 = model.predict(xtest1, verbose=0)

ignore = np.where(pred1.max(axis=1) < threshold)[0]
keep = np.where(pred1.max(axis=1) >= threshold)[0]
ni,nk = len(ignore), len(keep)

l,y = np.argmax(pred1[keep], axis=1), ytest1[keep]
cm, acc = ConfusionMatrix(l,y, num_classes=6)
```

Listing 5.2 Using test data to determine a softmax threshold

The process is linear: load the requested model (model), partition the test data (xtest0 and xtest1), gather softmax predictions and labels (pred and plabel) and the maximum softmax value for the correct (pc) and incorrect (pw) predictions.

The threshold is returned by the call to Search, which uses the maximum softmax values to select a threshold value that simultaneously maximizes the number of correct predictions above the threshold and the number of incorrect predictions below the threshold; see Listing 5.3.

```
def Search(pc, pw):
    thresh, mx = 0.1, -1.0
    for t in np.linspace(0.1, 1.0, 10000):
        nc = len(np.where(pc >= t)[0])
        nw = len(np.where(pw < t)[0])
        score = nc/len(pc) + nw/len(pw)
        if (score > mx):
            mx = score
            thresh = t
    return thresh
```

Listing 5.3 Brute-force search for a threshold

The search tests candidate thresholds via brute force running from 0.1 to 1.0 in 10,000 steps. The selected threshold maximizes score, which is the sum of the fraction of correct classifications above the threshold and the fraction of incorrect below the threshold. The idea is to pick a threshold that gives us the best chance of being correct when we decide to label inference-time samples with a maximum softmax value above it.

The remainder of *thresholds.py* (Listing 5.2) uses the threshold with the other half of the test set. Those classifications are divided into two sets, one to ignore because the maximum softmax value is below the threshold, and others to keep, as we have increased confidence in their possible correctness. The confusion matrix is calculated and then displayed.

Let's try *thresholds.py* for selected models trained on *bird6*. For example,

```
> python3 thresholds.py rgb 32 model.keras
Assigning labels to 115 predictions (0.58081)
(threshold=0.956886)
[[18  0  1  1  5  0]
 [ 0  4  0  3  0  3]
 [ 0  0  8  0  0  0]
 [ 0  0  4 11  0  4]
 [ 6  0  1  6 21  0]
 [ 0  1  0  4  0 14]]
Overall accuracy = 0.66087
Ignoring 83 predictions (0.41919)
```

This example uses a ResNet-18 model trained with 32×32 RGB images and 20× augmentation. Half the test set contains 198 images, of which 115 (58 percent) produced maximum softmax values above the threshold of 0.9569 with the remaining 42 percent ignored because their maximum was below the threshold.

The confusion matrix is calculated from only those deemed worthy giving us an overall accuracy of 66.1 percent. The model's accuracy without thresholding was 52.8 percent, meaning abstaining from a decision improves performance by about 13 percent. Naturally, abstaining comes with its own cost, namely, that of not making a decision, but it isn't difficult to imagine a scenario where such a decision is advantageous. If we begin to trust the model when making decisions above the threshold, we might save the time and effort otherwise spent analyzing or chasing down likely dead-end leads.

Applying *thresholds.py* to VGG8 and LeNet-5 models trained on an augmented 32 × 32 RGB dataset produces similar performance gains, but distinctly different threshold values:

```
Assigning labels to 125 predictions (0.63131)
(threshold=0.985599)
[[27  0  1  1  2  0]
 [ 0  5  0  0  0  4]
 [ 0  0  7  0  0  0]
 [ 3  0  5  8  3  4]
 [14  0  1  3 17  1]
 [ 0  0  0  2  0 17]]
Overall accuracy = 0.64800
Ignoring 73 predictions (0.36869)
```

for VGG8 and

```
Assigning labels to 109 predictions (0.55051)
(threshold=0.980108)
[[14  0  2  1  4  0]
 [ 0  7  0  0  0  4]
 [ 0  0  5  1  0  0]
 [ 2  0  2  8  6  1]
 [ 4  0  1  6 23  2]
 [ 0  0  0  3  0 13]]
Overall accuracy = 0.64220
Ignoring 89 predictions (0.44949)
```

for LeNet-5.

The VGG8 and LeNet-5 models produce consistent overall accuracies when they do decide on a class label. The VGG8 model uses a higher threshold than the others (0.9856) while also being willing to classify more inputs (63 percent). Of the three, LeNet-5 is the least willing to assign a label, and the least accurate when it does.

Accuracy improvements are not the same across classes. For instance, the problematic class 4 (Red-tailed Hawks) becomes 61.7 percent accurate when using the ResNet-18 model, but only 47.2 percent for VGG8, and improves the most to 63.8 percent with LeNet-5.

Building systems, especially those meant for use in (literally) the wild, such as monitoring birds, implies a higher likelihood of out-of-distribution inputs to the models – here's where decision thresholds come in handy, as we explore next.

5.3.2 Out-of-Distribution Inputs

A decision threshold gives us a tool we can use to detect at least some out-of-distribution inputs. To illustrate this requires some OOD images (not to be confused with the Ood). The file *build_ood.py* provides them for us to produce NumPy files in the *data* directory:

```
ood_32_xtrain.npy
ood_64_xtrain.npy
ood_gray_32_xtrain.npy
ood_gray_64_xtrain.npy
ood_ytrain.npy
```

These files mimic the various *bird6* datasets.

Specifically, we'll use images of Swainson's Hawks, Cooper's Hawks, Great Egrets, Say's Phoebes and Black-crowned Night Herons, along with dragonflies and wildflowers, as examples. Figure 5.1 presents examples of the OOD birds.

The OOD datasets include two hawks similar to red-tails, the Great Egret, which is quite similar to a Snowy Egret, and another heron to pair with the Great Blue Heron.

The file *inference.py* gives us what we need to pass the OOD datasets through existing models with an optional decision threshold. I leave perusing the file up to you; it follows the same pattern we used previously: load a dataset, preprocess it, run it through a model, and report the results.

Let's conduct a series of experiments using *inference.py* beginning with *bird6* data. The command line, one argument per line for easier review, is:

```
python3 inference.py
    ../data/bird6_gray_64_xtest.npy
    ../data/bird6_ytest.npy
    results/bird6_gray_64_f20_resnet18/model.keras
    0.995950
    tmp
```

The first two arguments are the source images, here 64 × 64 grayscale, and associated labels or none if there are none. The following argument is the name of a trained model, here a ResNet-18 trained on 20× augmented data. The threshold value comes next (0 implies none) followed by the name of an output directory, tmp.

The output, because test labels were provided, includes two confusion matrices: one without thresholding and one with,

```
Assigning 177 labels (0.44697) (threshold=0.995950)
No threshold:
[[61  1  5 10  6  2]
 [ 0 12  2  3  0 12]
 [ 1  0 19  2  1  0]
 [10  1  5 48  9  7]
 [39  5 15 28 42  4]
 [ 0  6  2  9  0 29]]
Overall accuracy = 0.53283
```

Figure 5.1 Top row (left to right): Swainson's Hawk, Cooper's Hawk, Great Egret. Bottom row: Say's Phoebe, Black-crowned Night Heron.

```
With threshold:
[[39  0  3  4  0  0]
 [ 0 10  0  0  0  7]
 [ 0  0 13  0  0  0]
 [ 3  0  1 27  2  2]
 [14  2  3 11 14  1]
 [ 0  2  0  3  0 16]]
Overall accuracy = 0.67232
```

In this example, thresholding improved the overall accuracy from 53.3 to 67.2 percent assigning labels to 45 percent of the input samples. The threshold of 0.995950 is quite high, but it is the threshold returned for this model (see *thresholds.py*).

If no test labels are given, as is the case for the OOD data, we get:

```
Assigning 105 labels (1.00000) (threshold=0.000000)
keep   : [11  8 13 40 15 18]  ( 105)
ignore: [0 0 0 0 0 0]  (   0)
```

without a threshold because we are forcing the model to select an output class always and

```
Assigning 32 labels (0.30476) (threshold=0.995950)
keep   : [ 2  0  4 12  3 11]  ( 32)
ignore: [ 9  8  9 28 12  7]  ( 73)
```

with a threshold.

There are no confusion matrices because there are no known class labels (everything's OOD). However, notice that without a threshold, all 105 inputs are dumped into one of the six classes even though none of them are members of that class. With a threshold, though, most of the inputs, $73/105 = 69.5$ percent, are ignored by the model, as we would hope. Curiously, most of the ignored OOD samples would have been declared of class Great Blue Heron. Why? Your guess is as good as mine. Perhaps there are strong, basic features the model associates with blue herons that are also present in many other input images.

Thresholding ignored some 70 percent of the OOD inputs, implying 30 percent were assigned a label. We must always be aware that models do at times make highly confident predictions for OOD data. Our simple threshold will not capture those errors.

A natural question to ask is: which OOD samples fall into which categories? In this case, we do have OOD labels, by design, allowing us to satisfy our curiousity. We have seven OOD classes placed into six categories by our models, implying a 7×6 "confusion matrix" where rows represent the OOD label and columns the model-assigned label. The file *ood_analysis.py* takes output produced by *inference.py* (no threshold) to print such a matrix. The script *ood_analysis_run* (apologies for the tedious names) tests RGB and grayscale 64×64 OOD images for each model type. For instance, the grayscale VGG8 model produced the matrix in Table 5.1.

Reviewing the table tells us that 10 of the Swainson's Hawks were classified by the model as Red-tailed Hawks. This makes sense given how close the two species appear visually. Indeed, this is a welcome outcome in that the model's confusion indicates that it did learn something about "hawkness," at least as far as larger hawks are concerned. The fact that most Cooper's Hawks were (strangely) assigned to the Great Blue Heron category argues somewhat against the model possessing a robust notion of "hawkness."

The other expected outcome is that virtually all Great Egrets fell into the Snowy Egret column. Again, this makes sense, as the two species are even more visually similar.

Table 5.1 shows counts, the label corresponding to the maximum softmax element for each OOD input. Missing are the maximum softmax values themselves, which might indicate the model's faith in particular class assignments. For example, are the Swainson's Hawks declared Red-tailed Hawks given high confidence, say well above any threshold determined by *thresholds.py*? If so, simple thresholding will not separate the two species.

For example, the 10 Swainson's hawks assigned to the red-tail category produced maximum softmax values of (truncating at three decimals):

$$0.964, \quad 0.998, \quad 0.999, \quad 0.995, \quad 0.999,$$
$$0.999, \quad 0.550, \quad 0.699, \quad 0.831, \quad 0.978$$

Table 5.1 Out-of-distribution *bird6* class assignments (VGG8, grayscale 64×64)

	Kestrel	Pelican	Kingfisher	Blue Heron	Red-tail	Snowy Egret
Cooper	1	3	2	7	2	0
Great Egret	0	1	0	2	0	12
Swainson	1	1	0	2	10	1
Night Heron	1	2	3	4	1	4
Phoebe	7	0	0	2	6	0
Dragonfly	0	1	6	5	0	3
Flower	0	6	4	3	0	2

The majority of these values are above 0.95, indicating the model is quite confident in its incorrect assignments. The situation for Great Egrets is even more extreme:

$$0.998, \quad 0.304, \quad 0.999, \quad 0.999, \quad 0.994, \quad 0.999,$$
$$0.999, \quad 0.999, \quad 0.681, \quad 0.416, \quad 0.999, \quad 0.999$$

The model's confidence on these OOD classes is a double-edged sword. On the one hand, it illustrates that the model has learned meaningful things about Red-tailed Hawks and Snowy Egrets. On the other hand, these results demonstrate that the model has yet to learn detailed features applicable to separating the two hawk and egret species. We won't do it here, but adding a *none-of-the-above (NOTA)* class containing Swainson's Hawks and Great Egrets would give the model examples of what are not Red-tailed Hawks and Snowy Egrets.

Assignment of the non-bird inputs is less straightforward. Dragonflies are split between kingfishers and blue herons while wildflowers remind the model of pelicans for some reason. This is an excellent time to remind ourselves that deep neural networks are essentially black boxes incapable of explaining their behavior.

Table 5.1 used no decision thresholding to force every input into a known category. The selected threshold for the VGG8 model was 0.993339. If a third argument to *ood_analysis. py* is given, and that argument points to the list of indices kept when applying a threshold (supplied by *inference.py*), we are arrive at Table 5.2 for the 36 percent of OOD inputs classified by the model.

Table 5.2 is relatively sparse, which we hope to see. There are 15 samples of each OOD class. Summing across the rows of the table indicates how many were still assigned a label with the worst offender the Great Egret (10 out of 15) and the least offensive the Cooper's Hawk (2 out of 15). The other OOD classes were roughly diminished to approximately the same degree. Naïvely, I expected the dragonflies and wildflowers to virtually go extinct, but they didn't. The inscrutable nature of neural networks likely means we will never know why.

This section introduced us to the concept of a decision threshold, which we can use to help improve the model's accuracy at the cost of refusing to make a decision for some fraction of inputs. The threshold was selected via the maximum softmax value using correct and incorrect classifications to decide on the transition point. A single threshold clearly works, but the threshold was chosen without regard to the assigned class label. It is possible to extend the approach to per-class thresholds. For example, suppose the input sample is assigned to class 3 because the largest softmax vector value is in the 4th position (recall, count from zero). In that case, that value can be compared to the threshold for class 3

Table 5.2 Out-of-distribution assignments when using a decision threshold

	Kestrel	Pelican	Kingfisher	Blue Heron	Red-tail	Snowy Egret
Cooper	0	0	0	2	0	0
Great Egret	0	0	0	1	0	9
Swainson	0	0	0	0	5	0
Night Heron	0	0	0	2	0	2
Phoebe	1	0	0	1	4	0
Dragonfly	0	0	3	1	0	1
Flower	0	3	1	2	0	0

to decide whether to keep the classification. It is entirely possible that class 3's threshold is somewhat different from, say, class 11's. I leave per-class thresholding as an exercise for the motivated reader.

Finally, the section demonstrated the utility of thresholding for detection (ignoring) of out-of-distribution inputs. Most readers will be interested in systems operating in the wild where there is little control over what might be acquired by the system, so OOD detection becomes of interest. Simple thresholding helps in this case, and I suspect that per-class thresholding would help even more.

Let's move on now to attempt to improve performance by utilizing the wisdom of the crowds.

5.4 Ensembling

Models have moods, in a sense. That is, they make mistakes in characteristic ways, just like people. The ad hoc notion of the "wisdom of the crowds" implies that even if individuals are prone to errors, an ensemble of individuals is less likely overall to make mistakes as each individual's mistakes are slightly different. The same notion applies to machine learning models, including neural networks. Let's test this by ensembling our models to combine their output in different ways with the hope of reducing error by playing to the strengths of each.

If you executed the *train_bird6_augment_models* script, you have LeNet-5, VGG8 and ResNet-18 models for every combination of image size and color depth. When used for inference, the models output a softmax vector from which we typically extract the index of the maximum and use that as the assigned label. Ensembling, in this instance, uses the softmax vectors from each model in the ensemble to produce a new prediction, either a new softmax vector from which the maximum value labels the input, or by voting on the model-assigned labels.

The file *ensemble.py* makes predictions from ensembles using one of three possible approaches: the arithmetic mean of the softmax vectors, the geometric mean of the vectors, or voting.

The arithmetic mean is nothing more than the average. If there are n classes in the dataset, then the softmax vector has n elements. If there are N models in the ensemble, the mean softmax vector is

$$\bar{s} = \frac{1}{N} \sum_{i=0}^{N-1} s_i$$

where s_i is the softmax output vector produced by model i for the current input to the ensemble. The assigned class label is the index of the maximum element in $\bar{s}$, as usual.

The geometric mean is the N-th root of the N softmax vectors multiplied together:

$$\bar{s} = \sqrt[N]{s_0 s_1 \cdots}$$

where multiplication is elementwise, not a vector dot product. The assigned class label is again the index of the maximum element in $\bar{s}$.

Finally, voting is a majority vote among the class labels selected by each model with ties chosen at random.

Run *ensemble.py* like so:

```
python3 ensemble.py ../data/bird6_64_xtest.npy
                    ../data/bird6_ytest.npy
                    avg
                    lenet5.keras vgg8.keras resnet18.keras
```

substituting the appropriate test dataset and labels along with trained model files as desired.

The example uses all three models trained on 64 × 64 RGB images and 20× augmentation along with softmax averaging to produce the ensemble output. My run generated:

```
[[59  1  1  7 17  0]
 [ 1 12  0  8  0  8]
 [ 0  0 20  3  0  0]
 [ 5  0  7 42 20  6]
 [34  0  7 21 67  4]
 [ 1  0  0  7  0 38]]
Overall accuracy: 0.60101
```

The ensemble's overall accuracy was 60.1 percent. The individual model's accuracies were 56.6 percent (LeNet-5), 56.8 percent (VGG8) and 57.8 percent (ResNet-18), respectively. Ensembling proves the wisdom of the crowds in this case producing a modest 2 percent increase in performance.

Averaging the softmax vectors before selecting the class label is only one of the three approaches supported by *ensemble.py*. The geometric mean and voting produced overall accuracies of 59.6 and 60.1 percent, respectively.

There is no free lunch. Ensembling comes with costs – increased inference time to pass the sample through multiple models, and increased storage requirements for model parameters.

The file *ensemble_test* trains ten LeNet-5 models on 64 × 64 RGB *bird6* images using 20× augmentation (minibatch 64, epochs 24). If the resulting models are all ensembled using *ensemble.py* (averaging), the overall test set accuracy is 55.1 percent. However, if the top three best-performing individual models are used, the accuracy jumps to 60.1 percent, matching the performance when ensembling LeNet-5, VGG8 and ResNet-18 models.

Training times differ significantly between the three architectures, implying that a viable option when considering ensembles is to train multiple simple models and then select top-performing models to ensemble together. In this case, the best single LeNet-5 model achieved an accuracy of 57.3 percent indicating that ensembling was still beneficial. Notice that multiple models using the same architecture were combined. Random neural network initialization ensures that each model is error-prone in (likely) different contexts, though disparate architectures are often more effective.

5.5 Discussion

What are we to make of our investigations with the *bird6* dataset? I think there are several key points to consider, especially when building new datasets that are likely to be on the smaller side (which seems to be most in practice). Five areas come to mind: hyperparameter tuning, data augmentation, decision thresholds, out-of-distribution detection, and ensembling.

5.5.1 Hyperparameter Tuning

Hyperparameter tuning is essential with small datasets. There's often a sweet spot for minibatch size and number of training epochs. It's difficult with a small dataset to hold some back for validation to guide the training process. Options in this area that we didn't explore directly, but which are straightforward to add to the existing set of experiments, include *k-fold cross validation* and *early stopping*. With k-fold cross validation, the existing train and test data, merged, is split into k equal-sized chunks (the "folds") with each fold successively held out as test data when training the model on the remaining $k - 1$ folds. The performance of each fold as test is then averaged to understand how well the full model trained on all data might perform in the wild. If that average level of performance is adequate, train on all available data, then deploy the model. Augmentation is possible if applied separately for each tested fold so that the current training set and current test set do not overlap (i.e., the augmented version of one does not appear in the other).

Early stopping means adding a callback routine to the training process (in Keras) that monitors the validation loss and stops after a set number of epochs have failed to reduce the loss. After stopping, the weights that best perform on the validation set are returned as the trained model.

Cross validation produces k separately trained models. *Cross validation ensembling* uses each model as a member of an ensemble, sometimes including the final model trained on all the available data for an ensemble of $k + 1$ models. The random nature of neural network initialization and training implies that the models will be much like the multiply trained LeNet-5 models in terms of errors so that an ensemble can be expected to perform better than any individual model.

5.5.2 Data Augmentation

If you lack data, acquire more. That's easy to state but often difficult to do. In my experience, practical deep-learning projects are perpetually data-starved. Therefore, if real data is unavailable, invent some via data augmentation.

We explored data augmentation techniques for images but we barely scratched the surface of what is possible. More advanced data augmentation alters different features of the images, like shading, color (gamma) and backgrounds. Even seemingly bizarre alterations are at times helpful. What models use as visual clues is not necessarily what humans use.

Data augmentation is the most straightforward regularizer after simply acquiring more real data. It should be used whenever possible, but with care to ensure that augmented versions of training samples do not end up in the test set (yes, this has happened more than once in practice).

The advent of generative AI, particularly diffusion methods that create images from text prompts, offers a potentially novel approach to data augmentation, with suitable cautionary warnings in place.

For example, Figure 5.2 presents a true Say's Phoebe image (upper left) along with seven re-imagined versions courtesy of Stablity AI's DreamStudio. The first four are based on the original, and the final three (lower right) are produced from just a text prompt.

Figure 5.2 Say's Phoebe (upper left) and seven re-imagined versions using generative AI.

Are these re-imagined images suitable as augmented versions of the original? Testing is required to know better. Of course, the generated versions can themselves be augmented by traditional means.

5.5.3 Decision Thresholds

The standard approach selects the greatest softmax output element and assigns its index as the class label. Models trained on small datasets are often of marginal utility and it becomes advantageous to designate a decision threshold based on the largest softmax value. If the sample under consideration produces a maximum softmax element below the threshold, refuse to assign a label and, optionally, pass the sample off to another classifier or human for further analysis.

We experimented with this approach using a single decision threshold applied to all classes and learned that it was helpful. A natural extension determines per-class thresholds assuming sufficient representation of the class in the training set. I leave that as an exercise for the reader. When using per-class thresholds with augmented *bird6*, do matters improve?

5.5.4 Out-of-Distribution Detection

Thresholding makes it possible to detect out-of-distribution inputs to the model, at least some of the time. If the model is deployed in a scenario where the set of possible inputs is relatively unconstrained, then it makes sense to develop a set of OOD inputs and use them to select decision thresholds that provide strong evidence that the input is OOD. Note that such thresholds might differ from the "yes, I want to assign a label" threshold. For example, if θ_{label} is the threshold above which we will assign a label and θ_{ood} is the threshold below which we insist on calling the input OOD ($\theta_{OOD} < \theta_{label}$), then the maximum softmax vector element, θ, falls into one of these three categories:

$$\theta \geq \theta_{label} \tag{a}$$

$$\theta_{OOD} \leq \theta < \theta_{label} \tag{b}$$

$$\theta < \theta_{OOD} \tag{c}$$

If (a), label the input as we believe the model's label stands a good chance of being correct. If (c), label the input OOD for the same reason. Samples landing in (b) are beyond the model's abilities and should be segregated for further analysis.

As previously mentioned, θ_{label} and θ_{OOD} can be selected on a per-class basis with the OOD threshold based on the class label into which particular OOD examples fall during model testing.

5.5.5   Ensembling

The mistakes a model makes are characteristic of the model's parameters, which themselves are a consequence of the model's initialization and training process. Therefore, it seems sensible at times to ensemble multiple model outputs to arrive at a better decision for the assigned class label. Ensembling is especially helpful with small models trained on small datasets.

Our experiments with *bird6* showed this to be the case. Specifically, we learned that test set performance improved when combining the output of LeNet-5, VGG8 and ResNet-18 models. We also discovered that training a bigger set of simple LeNet-5 models then ensembling the best three produced an effectively identical performance increase, implying the ensembles of simpler models can achieve good utility when carefully selected, provided the computational cost inherent in assessing inputs with multiple models is acceptable.

The experiments of this chapter give us a suite of tools to use when working with smaller datasets. However, in general, we can often do better by beginning with a large model pretrained on a large dataset, even if the dataset is not precisely of the kind we want. The next chapter demonstrates how.

6. Using Pretrained Models

Chapter 4 introduced the *bird6* dataset. Chapter 5 saw us apply common deep learning techniques to improve the performance of models trained on the limited *bird6* dataset, including data augmentation, ensembling, and decision thresholds. Augmentation delivered a best-performing model (ResNet-18, RGB, 64 × 64) with an overall test set accuracy of just below 57 percent compared to a random guess accuracy of about 17 percent. Ensembling the three model types explored (LeNet-5, VGG8 and ResNet-18) bumped the accuracy to 60 percent. In this chapter, we explore additional approaches that have become standard fare in the deep learning community: transfer learning and fine-tuning. Both approachs rely on the existence of large, pretrained models, known as *foundation models*. Such models have already learned essential features present in many natural images, especially photographs. Foundation models are trained on massive collections of labeled images, often the 1.2 million present in the 1,000-class ImageNet dataset, including 59 types of birds (see the file *ImageNet_birds.txt*). The chapter's structure is straightforward. First, we dive into the concepts behind transfer learning and fine-tuning. Then, we build on that framework with experiments to give us the experience we need to apply these techniques, especially to projects related to smaller datasets. As we'll come to learn, it's possible to improve substantially on the 60 percent accuracy of Chapter 5, mainly by turning to powerful multimodal models trained jointly on images and text.

6.1   Understanding Transfer Learning and Fine Tuning

In a nutshell, *transfer learning* uses a pretrained model to produce output embedding vectors as a new representation of input images, then uses those new representations for downstream tasks including training other machine learning models. *Fine tuning*, on the other hand, adds a task-specific head to a pretrained model. Then, with optional freezing of specific existing base model layers, trains the entire model on a new dataset in the hopes that the feature representations learned by the base model provides a good starting point leading to rapid convergence on the new dataset.

Figure 6.1 illustrates transfer learning. The top part of the figure shows the first step: training the base (foundation) model. ResNet-50, a 50-layer version of the ResNet-18 model we used in earlier chapters, is indicated, and is one of three foundation architectures we'll explore in this chapter. The second is known as MobileNet. The third, CLIP, is a breakthrough model from OpenAI that has jointly learned imagery and text.

Transfer learning's first step produces a trained foundation model. The computation necessary to train the model is often extensive, orders of magnitude beyond what we encountered in Chapter 5, even with data augmentation. However, training a foundation

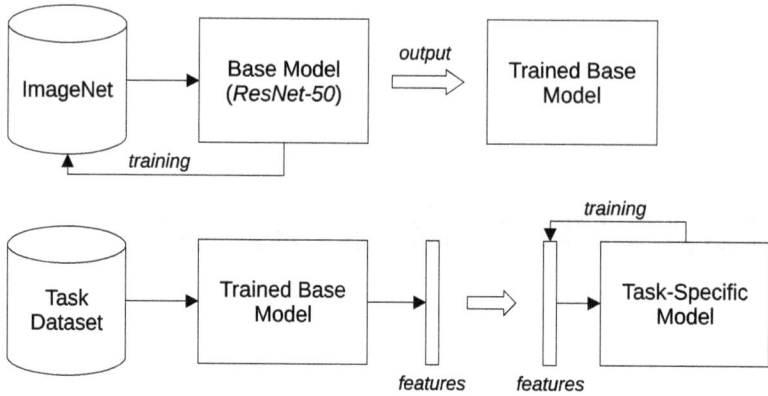

Figure 6.1 Transfer learning: training the foundation model (top) and using task-specific features derived from it (bottom).

model like ResNet-50 or MobileNet on the ImageNet dataset is child's play compared to the level of compute necessary to train foundation LLMs like ChatGPT or Claude, etc. Such LLMs contain over a trillion parameters and use training sets typically characterized as "the entire Internet." It's no wonder that companies like Microsoft are seeking to build nuclear reactors to power their data centers.

Step 1 produces a trained base model. Step 2, the lower part of Figure 6.1, strips the base model's head (the topmost layer producing softmax outputs) making the last dense layer, or equivalent before the softmax layer, the model's new output.

The new output is a one-dimensional *embedding vector* representing the model's internal characterization of an input image. Buried (embedded) within this vector is everything the model learned about the input as it flowed through the network, layer by layer. In a sense, the base model "transmogrifies" its inputs to turn them into an information-rich version that, hopefully, captures essential characteristics of the task dataset (with suitable apologies to Watterson). I'll refer to embedding vectors as feature vectors from time to time to emphasize how we intend to use them.

Step 2 uses the *unmodified* weights and biases of the base model to produce embedding vectors, one for each sample in the task dataset. If the task dataset is similar to the dataset on which the base model was trained, then it is reasonable to expect the resulting embedding (feature) vectors to be similarly information-rich and, therefore, to be useful for downstream tasks, including using the feature vectors as training data for a simpler model, often a traditional machine learning model like a random forest, support vector machine or multilayer perceptron.

The critical statement in the preceding paragraph is, "the task dataset is similar to the dataset on which the base model was trained." If the data is radically different, like magnetic resonance images, passing it through a base model trained on natural photographs, like ImageNet, will give us little reason to hope that the resulting embedding vectors will be information-rich. They might be helpful, but then again, they might not. Fortunately, we intend to pass bird images through a base model trained on ImageNet, so we expect the resulting embedding vectors to be meaningful.

Transfer learning seeks to use the compressed, information-rich embeddings of the base model to make training a simpler machine learning model more successful than training from scratch. Transfer learning does this by using the base model for inference only, avoiding computation to adjust its weights and biases.

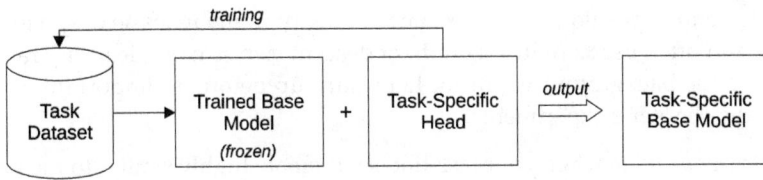

Figure 6.2 Fine tuning retrains the base model on the task dataset. Note that the base model weights are often (partially) frozen.

Simple machine learning models often do best with small datasets, implying that feature vectors derived from a base model are likely already sufficient for training a simple model. Further, training a simple model means little computational overhead, at least compared to the compute that went into training the base model. Therefore, favor using transfer learning when the task dataset is small, of similar character to the dataset on which the base model was trained, and you have limited compute resources for training a simpler machine learning model on the resulting embedding vectors.

Fine-tuning retains Step 1 of Figure 6.1, but replaces Step 2 with the contents of Figure 6.2 to construct a refined, task-specific version of the base model.

The weights and biases of the base model become the initial state of the task-specific model, which is then trained on the small to moderate-sized task-specific dataset. Recall that the base model's head is removed in transfer learning to produce embedding vectors. Fine-tuning adds a new head, one suited to the number of classes in the task dataset retaining the base model as the feature generator. Adding a head to the base model before training both parts jointly helps to condition the entire model to the task data, which is helpful if the task dataset is of a kind somewhat different in nature (meaning statistical properties) from the dataset on which the base model was trained.

The base model in Figure 6.2 is marked as "frozen," implying that its weights and biases are locked and not allowed to change during training. If the entire base model is frozen and only the new topmost layers are altered by training, we are in a situation functionally equivalent to using embedding vectors with a classic machine learning model, especially an MLP. However, if some of the base model layers are allowed to change during fine-tuning, the opportunity exists to push the entire model into a better location in the error space, thereby producing a new version of the base model better suited to the task.

A question naturally presents itself: which layers of the base model should be frozen and which should be allowed to adapt during fine-tuning? There is no hard and fast rule regarding freezing and unfreezing beyond a general observation that the lower layers of a CNN adapt to basic image features like edges, textures and colors, while a higher level adapts to larger groupings of features related to the actual classes on which the model is trained. Therefore, it is often observed that freezing low-level convolutional layers is helpful, but allowing the very topmost layers of the base model to adapt during fine-tuning produces better outcomes. Of course, deep learning is a strongly empirical science, implying per-dataset variation in this observation.

Fine-tuning assumes that the base model is already conditioned to the task, at least partially; therefore, training with the task-specific dataset should be kept to a minimum to prevent overfitting. Conceptually, the base model plus task-specific head contains a collection of weights and biases that are "close" in error space to a reasonable minimum (or saddle point), and fine-tuning will likely need only a handful of gradient descent steps to arrive at a refined position. Therefore, fine-tuning typically relies on only a few epochs of training and often uses a reduced gradient descent step size (i.e., learning rate).

We have the background we need. Let's sum up before an important caveat and beginning the chapter's experiments:

1. Use *transfer learning* when your task dataset is small, highly similar to the dataset on which the base model was trained and you have limited computational resources.
2. Use *fine tuning* when your task dataset is of moderate to large size, reasonably similar to the dataset on which the base model was trained and you have access to sufficient computational resources.

And now the caveat: we'll ignore fine-tuning henceforth. Why? As we'll learn during the experiments, for bird images, transfer learning with pretrained models, especially newer models like CLIP, essentially removes the need for fine-tuning for most practical situations.

We'll proceed like so: first, we'll build a "foundation" model using an external bird dataset and then experiment with it to understand how it may (or may not) improve our *bird6* results. Second, we'll replace the custom transfer learning model with ResNet-50 and MobileNet models pretrained on ImageNet. Third, we'll substitute CLIP for the pretrained vision models to learn what a state-of-the-art embedding model is capable of. Fasten your seatbelts; we're about to take off.

6.2   Using Birds 25

We have the *bird6* dataset. It's specific to the task at hand, but not particularly extensive in scope. The techniques of Chapter 5 only go so far, which is why we are now turning to transfer learning. We need a base model, a model trained on a massive dataset that isn't our task dataset but is reasonably related to it in terms of data characteristics. Such models exist and are freely available for anyone to use.

We begin, however, rather modestly. We're interested in bird images, so working with a base model trained on the same is sensible. We'll trade the desire for a base model trained on a "massive" dataset for a modest one trainable in a reasonable amount of time. There is such a bird dataset, the Birds 25 dataset. We'll use this dataset to train base models conditioned to detect birds, then use those models to generate embedding vectors for the *bird6* images. The Birds 25 "foundation" models are merely pedagogical; don't get your hopes up too high.

6.2.1   Building the Birds 25 Models

The Birds 25 dataset is in the public domain and may be downloaded from Kaggle (free account required):

```
www.kaggle.com/datasets/ichhadhari/indian-birds
```

The processed NumPy files are also part of the book's data package available on my website; see the Introduction.

The Birds 25 dataset consists of 30,000 images of 25 different Indian birds. The file *Birds_25.txt* lists the represented species. We'll use the images in this dataset to build a base model that (we hope!) is sufficiently well versed in detecting "bird-ness" that we can use the embeddings it generates to train simpler machine learning models on *bird6*.

The file *build_birds_25.py* processes the raw Birds 25 images. I recommend working with the resulting NumPy files acquired via the book's data package, but if you do register with Kaggle, you'll have access to not only Birds 25 but a vast number of other datasets and models.

The goal of *build_birds_25.py* is to transform the raw Birds 25 train and test images into a collection of 64×64 RGB images. We'll focus on color imagery in this chapter. Straightforward modifications to the code will produce grayscale versions, should you wish them.

The code performs the necessary file gymnastics to access the train and test images (called "valid" in the dataset) before handing each to the Chip function to extract a central square region that is resized to 64×64 pixels:

```
def Chip(fname):
    img = np.array(Image.open(fname).convert("RGB"))
    h,w = img.shape[:2]
    if (h > w):
        lo,hi = (h-w)//2, (h-w)//2 + w
        im = img[lo:hi, :, :]
    else:
        lo,hi = (w-h)//2, (w-h)//2 + h
        im = img[:, lo:hi, :]
    img = Image.fromarray(im).resize((64,64), Image.BILINEAR)
    return np.array(img)
```

The image's aspect ratio decides which central portion is extracted. Note that the largest such square chip is selected before resizing to 64×64 pixels. The code generates the expected set of NumPy files:

```
birds_25_xtrain.npy
birds_25_ytrain.npy
birds_25_xtest.npy
birds_25_ytest.npy
```

If you examine the images, you'll notice that training images are sometimes flipped upside down. I find this curious, after all, there are no hummingbirds in the class list, but it isn't necessarily detrimental to the model as doing so implicitly conditions the model to notice features relevant to "bird-ness" as opposed to features related to the standard orientation of the bird. In other words, the model is pushed to develop detection kernels tuned to the features associated with the birds themselves, and less so their immediate surroundings.

Let's train two base models, a VGG8 and a ResNet-18. The necessary code is in *pretrain.py* and it follows the same sequence we've used previously: load the Birds 25 data, preprocess as needed, build the requested model and train for the given number of epochs before storing the model on disk. For example, the VGG8 model was produced via:

```
> pretrain.py vgg8 64 20 results/vgg8_birds_25
```

with a similar command line for ResNet-18. Notice that the base models are trained for 20 epochs using a minibatch of 64. The resulting models are reasonably accurate on the Birds 25 test set:

```
Test set accuracy: 0.8289, MCC: 0.8221 (VGG8)
Test set accuracy: 0.8587, MCC: 0.8530 (ResNet-18)
```

The trained base models are included in the book's data package as:

```
vgg8_birds_25.zip
resnet18_birds_25.zip
```

Unzip them in the Chapter 6 source directory to extract the resulting .keras model files.

The two base models are working as we expect. Let's apply them in a few transfer learning experiments.

6.2.2 Experiments

Extracting embedding vectors from the VGG8 and ResNet-18 models requires manipulation to remove the existing classification heads. Once removed, we can pass new inputs through the model to generate the necessary embeddings.

The file *transfer.py* contains the code for our experiments. Much of the code is familiar, especially the functions to augment the *bird6* images. The *bird6* images are loaded, augmented and scaled [0, 1] as before:

```
x_train, ytrain = AugmentDataset(x_train, ytrain, factor)
x_train = x_train.astype('float32') / 255
x_test = x_test.astype('float32') / 255
```

Next, comes the desired pretrained network (mname is read from the command line):

```
if (mname == 'vgg8'):
    base = load_model('birds_25_vgg8.keras')
else:
    base = load_model('birds_25_resnet18.keras')
```

The base model must be wrapped in a new Model to extract the output from the second to last Dense layer:

```
model=Model(inputs=base.input, outputs=base.layers[-3].output)
```

Keras stores the network's layers in layers. The [-3] index refers to the layer just before the final Dense and Softmax layers. The new model, therefore, accepts the input used by the base model and outputs the specified layer, the embedding we want.

For example, the final few VGG8 layers (courtesy of `base.summary()`) are:

```
dense_1 (Dense)            (None,  2048)
batch_normalization_7      (None,  2048)
re_lu_7 (ReLU)             (None,  2048)
dropout_1 (Dropout)        (None,  2048)
dense_2 (Dense)            (None,  25)
softmax (Softmax)          (None,  25)
```

The layer name is on the left with the number of elements in the layer on the right (the `None` is from the number of values in the minibatch, which isn't specified by `summary`). Counting backward from −1 referencing the Softmax layer indicates that for VGG8, the embeddings come from the output of the Dropout layer, which is nothing more than the final Dense layer's output passed through batch normalization and ReLU. Recall that the Dropout layer isn't active during inference, which is how we use the pretrained base model. Therefore, the VGG8 embeddings are 2048-element vectors.

For ResNet-18, we get:

```
global_average_pooling2d    (None,  512)
dense (Dense)               (None,  25)
softmax (Softmax)           (None,  25)
```

telling us that the ResNet-18 embeddings are 512-element vectors, the output of the global average pooling layer.

To generate the embeddings requires passing the train and test data through the new model:

```
xtrn = model.predict(x_train, verbose=0)
xtst = model.predict(x_test, verbose=0)
xtrn = (xtrn - xtrn.min()) / (xtrn.max() - xtrn.min())
xtst = (xtst - xtst.min()) / (xtst.max() - xtst.min())
```

The last two lines rescale the embedding vectors so that every element lies within [0, 1]. This step isn't strictly necessary for all downstream uses, but some classical machine learning models, like MLPs, work best when the feature elements are all within the same range. Others, like random forests, don't care since features are evaluated in isolation.

At this point, `xtrn` and `xtst` are the augmented *bird6* images as understood by the selected, pretrained base model, which has transformed the raw image into a 2048- or 512-element, information-rich feature vector.

Let's define and train the simpler model. The code in *transfer.py* does this three times: first with the embeddings just derived, second with embeddings extracted from *untrained* VGG8 or ResNet-18 models, and, third, with the augmented *bird6* images themselves unraveled into 1-dimensional vectors. I'll describe only the first; the rest follow naturally:

```
if (mtype == "rf"):
    clf = RandomForestClassifier(n_estimators=opt)
```

```
else:
    clf = MLPClassifier(hidden_layer_sizes=(opt,opt//2),
                        max_iter=1000)
clf.fit(xtrn, ytrain)

pred = clf.predict(xtst)
cm,acc = ConfusionMatrix(pred, ytest, num_classes=num_classes)
mcc = matthews_corrcoef(ytest, pred)
print(cm)
print("Test set accuracy: %0.4f, MCC: %0.4f" % (acc,mcc))
```

The first code paragraph creates the desired classifier object. The variable `opt` is an integer read from the command line. If a random forest is created, `opt` specifies the number of trees in the forest. For an MLP, `opt` defines the number of nodes in the first hidden layer with half that number automatically used for a second hidden layer.

The second code paragraph uses the trained model (`clf`) to make predictions on the embeddings of the test set. The confusion matrix, overall accuracy and MCC are calculated and displayed. The same is displayed for embeddings derived from untrained base models and from a classifier trained directly on the images.

It makes sense to explore how well the embeddings from the Birds 25 pretrained models work in helping a simple classifier, and it's sensible to compare those results with the classifier's attempt on the raw images, but why explore embeddings derived from untrained base models?

Embeddings are transformations from the space of the inputs (images) to a new space, that of the embedding vectors (2048 or 512 dimensions). A randomly initialized CNN is effectively a random mapping from images to the embedding space. It's known that such random mappings often produce structure in the embedding space, just by chance, that is more easily utilized by a model than the highly correlated features found in an unraveled image. Therefore, an uninitialized base model might be expected to be somewhat successful on its own. If pretraining on Birds 25 adds nothing, we should expect the classifiers trained on those embeddings to be no better than classifiers trained on random embeddings. The code in *transfer.py* tests this hypothesis.

Let's take *transfer.py* out for a spin:

```
> python3 transfer.py mlp 512 vgg8 1
```

The command trains an MLP with 512 nodes in the first hidden layer and 256 in the second using VGG8 embeddings of *bird6* images without augmentation (the final `1` on the command line). Recall that we're working with 64×64 RGB images and that the VGG8 embeddings have 2048 elements, meaning the MLP maps a 2048-element input to 512 nodes to 256 nodes to 6 outputs (softmax).

Three confusion matrices and accuracies are ultimately displayed. The first shows the MLP's performance with the Birds 25 pretrained embeddings:

```
Embeddings (VGG8): (MLP 512)
[[46  5 12  7 14  1]
 [ 0 13  3  7  0  6]
 [ 0  0 20  1  2  0]
```

```
[10   0 25 33   9   3]
[41   3 12 10  61   6]
[ 0   2  3  4   2 35]]
Test set accuracy: 0.5253, MCC: 0.4234
```

The overall accuracy of 52.5 percent is a significant improvement over the 39.7 percent we found in Chapter 4 when training VGG8 from scratch on the same 413 images. Only when we augmented the training set 20× in Chapter 5 did we achieve a similar performance level, 56.7 percent. In other words, pretraining VGG8 on Birds 25 produced a model whose embeddings contain sufficient knowledge of "bird-ness" that a simple MLP can capture the essence of *bird6* data without augmentation – well, at least some of it.

The next confusion matrix indicates how well the randomly initialized VGG8 model's embeddings did when passed to the simple MLP:

```
Random mapping:
[[50  1  6  8 19  1]
 [ 0 14  2  4  0  9]
 [ 5  0 11  0  6  1]
 [ 7 11 17 16 21  8]
 [31 14 16 13 56  3]
 [ 1 14  1  5  0 25]]
Test set accuracy: 0.4343, MCC: 0.3041
```

The MLP was only 43.4 percent accurate, considerably less so than the model using the pretrained embeddings. This result is in line with my earlier claims that random projections might introduce structure exploitable by a model.

The final confusion matrix presents results from training the MLP on the raw images unraveled into feature vectors ($64 \times 64 \times 3 = 12, 288$ elements):

```
Unraveled images:
[[50 0  5 15  8  7]
 [ 0 2  0 20  0  7]
 [ 6 0 11  6  0  0]
 [13 0  7 45  3 12]
 [58 0 18 32 20  5]
 [ 0 1  0 21  0 24]]
Test set accuracy: 0.3838, MCC: 0.2575
```

Interpret this confusion matrix as what might have been found if this experiment had been run in the 1980s or early 1990s. There are six classes, so the result is well above the random guess rate of about 17 percent, but it is also some 14 percentage points below the pretrained model embeddings result. Notice, also, the abysmal performance on the Red-tailed Hawk class, the second to last row of the confusion matrix, with only 20 of the 133 red-tails in the test set correctly classified (15 percent). The MLP using pretrained features was 46 percent accurate on red-tails, which isn't anything to write home about but is a significant improvement over 15 percent.

Running again to train a random forest with 200 trees on VGG8 features produces:

```
Test set accuracy: 0.3687, MCC: 0.3020 (embedding)
Test set accuracy: 0.3131, MCC: 0.2059 (random)
Test set accuracy: 0.3535, MCC: 0.2338 (unraveled)
```

The random forest results are inferior to the MLP results, and quite variable if run multiple times. Random forests partition the data along feature dimensions and therefore cannot capture subtle, yet important, characteristics that do not align with any particular dimension of the input feature vector. This observation implies continuing with top-level MLPs for the remainder of the chapter. I left the random forest option in the code if you wish to experiment with it.

The original *bird6* training set, unaugmented, produced an MLP with an accuracy of 52.5 percent. It's natural to wonder how things will improve if we augment the *bird6* images and then pass them through the pretrained model before training the MLP. Fortunately, that's easy to test by changing the final 1 on the command line to 20:

```
> python3 transfer.py mlp 512 vgg8 20
```

The resulting output is no better than passing the unaugmented data through the pretrained VGG8, at least for the models trained on pretrained or random embeddings:

```
Test set accuracy: 0.5278, MCC: 0.4202 (embedding)
Test set accuracy: 0.4646, MCC: 0.3424 (random)
Test set accuracy: 0.4268, MCC: 0.3078 (unraveled)
```

The accuracy of the model trained on unraveled images did improve, from 38.4 percent to 42.7 percent, as we should expect for the same reason that training VGG8 from scratch with augmented images improved the model.

Augmenting the *bird6* dataset and then generating embeddings to train a simple MLP produced a model that is essentially identical to the model learned from the original dataset embeddings. In other words, the augmented images produced embedding vectors that were sufficiently similar to the original data embeddings so as to add virtually nothing in terms of diversity from which the MLP could produce a more accurate model.

At first glance, this might be somewhat confusing. After all, augmenting the images and then training VGG8 from scratch considerably improved performance. However, pretraining VGG8 on Birds 25 has likely already conditioned the model to be insensitive to basic data augmentations such as rotations, flips and shifts. Therefore, the output embeddings are essentially the same as if no augmentations were performed.

The script *transfer_test* executes *transfer.py* six times each for VGG8 and ResNet-18 features to train an MLP with 512 first hidden layer nodes without augmentation. The results, captured in *transfer_test_results.txt*, are parsed by *transfer_test_results.py* to produce a table comparing the means of each combination; see *transfer_test_summary.txt*. Table 6.1 contains the overall accuracies.

An examination of Table 6.1 demonstrates that the ordering between models trained on pretrained embeddings, random embeddings from the base model, and the unraveled images themselves for VGG8 preserves our earlier observation:

embedding > random > unraveled

Table 6.1 Overall accuracies and MCC (mean ± SE, $n = 6$).

	Accuracy	MCC
VGG8:		
embedding	0.52780± 0.00427	0.42927± 0.00438
random	0.41412± 0.01025	0.30212± 0.00980
unraveled	0.35478± 0.01833	0.23840± 0.01492
ResNet-18:		
embedding	0.43013± 0.00378	0.30707± 0.00407
random	0.26808± 0.01937	0.16002± 0.01431
unraveled	0.29882± 0.03414	0.20088± 0.02281

However, the order is altered for models using ResNet-18 features:

$$embedding > unraveled > random$$

implying that untrained ResNet-18 embeddings are less feature-rich than those from untrained VGG8 models.

Exactly why this is the case isn't immediately clear, but I suspect it is because ResNet models include skip connections enabling residual layers to learn offsets added to the layer inputs. When untrained, it seems reasonable to believe that random residual output dilutes any "signal" from the input image. VGG8, on the other hand, builds new features, layer by convolutional layer, that are more likely to contain structure that a top-level model can exploit.

The results of this section are sensible, but the fact that models pretrained on Birds 25 do not lead to better results than training a VGG8 from scratch with augmentation hints that pretrained models should probably rely on massive training sets. Fortunately for us, such massively pretrained models are readily available.

6.3   Using ResNet-50 and MobileNet

We lack sufficient data to train large models. However, we do, via Keras, have easy access to large models pretrained on the 1 million plus images in the ImageNet database. Let's use two such models to produce embeddings for transfer learning experiments: ResNet-50 and MobileNet.

We discussed the ResNet architecture in Chapter 2; see Figure 2.7 for a summary of the ResNet-18 architecture. The ResNet-50 architecture follows the same form but consists of 50 residual bottlenecks, not 18. MobileNet, on the other hand, uses a different structure and replaces two-dimensional convolutional layers with what are known as *depthwise convolutions*. MobileNet pairs depthwise convolutions with pointwise convolutions to produce a deep model that uses significantly fewer weights compared to other models, hence "mobile" in the name: the model is small enough to implement on a device like a mobile phone. Depthwise and pointwise convolutional layer operation is beyond what we need to consider here. You'll find more information in the resources of Chapter 11, should curiosity strike.

Using these models requires two steps. Step 1 accesses the models, pretrained on ImageNet, from within Keras to generate embeddings of *bird6* images. Step 2 then uses the embeddings to train simpler models in the hopes of producing a final model outperforming those of Chapter 5.

6.3.1   Creating *bird6* Embeddings

The file *imagenet_features.py* creates ResNet-50 or MobileNet models, loads pretrained weights for ImageNet before passing *bird6* images, optionally augmented, to produce output NumPy files containing embedding vectors. Let's extract the features before reviewing the code. For ResNet-50 features, use:

```
> python3 imagenet_features.py resnet 1
          results/features_resnet50
```

and

```
> python3 imagenet_features.py mobile 1
          results/features_mobilenet
```

for MobileNet. Notice that both sets of features use no augmentation (1×).

The command lines produce files like *xtrain.npy* and *xtest.npy* with associated label files in their respective directories. Each row of the files contains an embedding vector, the output of the pretrained model for the given input image. For example,

```
>>> import numpy as np
>>> np.load("results/features_mobilenet/xtrain.npy").shape
(413, 960)
>>> np.load("results/features_mobilenet/xtest.npy").shape
(396, 960)
>>> np.load("results/features_resnet50/xtrain.npy").shape
(413, 2048)
>>> np.load("results/features_resnet50/xtest.npy").shape
(396, 2048)
```

where we see that MobileNet produces 960-element embedding vectors while ResNet-50's vectors have 2048 elements. The simple models of the next section use these vectors in lieu of the raw bird images.

Let's review some of the code in *imagenet_features.py* beginning with necessary imports and creating the pretrained base model. We'll skip the image augmentation functions, they are identical to those of Chapter 5; see Listing 6.1.

```
import tensorflow as tf
from tensorflow.keras.applications import ResNet50,
          MobileNetV3Large
from tensorflow.keras.applications.resnet50
          import preprocess_input as rpreprocess
from tensorflow.keras.applications.mobilenet_v3
          import preprocess_input as mpreprocess
```

```
x_train = np.load("../data/bird6_64_xtrain.npy")
x_test = np.load("../data/bird6_64_xtest.npy")
ytrain = np.load("../data/bird6_ytrain.npy")
ytest = np.load("../data/bird6_ytest.npy")
x_train, ytrain = AugmentDataset(x_train, ytrain, factor)

if (mname == 'mobile'):
    model = MobileNetV3Large(weights='imagenet',
                include_top=False, pooling='avg')
    preprocess = mpreprocess
else:
    model = ResNet50(weights='imagenet',
                include_top=False, pooling='avg')
    preprocess = rpreprocess
```

Listing 6.1 Preparing data and pretrained models

First come the necessary imports from Keras. Pretrained models are available in the `applications` module and consist of a model class and an associated `preprocess_input` function. Each model includes a function with the same name, so we must create aliases to ensure the proper function is called. The preprocessing function performs whatever image manipulations are necessary to make the input match what the base model is expecting. Notice that we assign the proper preprocessing function to the variable `preprocess`. Python treats functions as first-class objects.

Next, the *bird6* train and test images are loaded and augmented as desired (`factor` is supplied on the command line).

The base model is then created by calling the appropriate constructor. Here's where we inform Keras that we want to use pretrained ImageNet weights and that we do not want the model to include a classification head but instead to replace the classification head with an average pooling layer. The output of the average pooling layer becomes the embedding vector representing the input.

We're now ready to generate the train and test embedding vectors:

```
xtrn = []
for i in range(len(x_train)):
    im = tf.image.resize(x_train[i], (224,224))
    im = preprocess(im)
    im = tf.expand_dims(im, axis=0)
    embedding = model.predict(im, verbose=0)
    xtrn.append(embedding.squeeze())
xtrn = np.array(xtrn)
```

A similar set of steps creates the test set vectors.

ImageNet uses 224×224 pixel RGB images, therefore, we must resize the 64×64 bird images accordingly. This step is only necessary if using pretrained weights. If we desire to train the ResNet-50 or MobileNet models from scratch, we can use the images as they are so long as they are at least 32×32 pixels.

The resized image is passed to `preprocess`, which contains the appropriate preprocessing function. Next, the image is expanded to match the expected shape of the input to the model before producing the embedding by calling `predict`. The variable `xtrn` accumulates the embeddings, image by image.

Finally, we rescale the vectors to [0, 1] and write them to disk for later use:

```
xtrn = (xtrn - xtrn.min()) / (xtrn.max() - xtrn.min())
xtst = (xtst - xtst.min()) / (xtst.max() - xtst.min())

np.save(outdir+"/xtrain.npy", xtrn)
np.save(outdir+"/ytrain.npy", ytrain)
np.save(outdir+"/xtest.npy", xtst)
np.save(outdir+"/ytest.npy", ytest)
```

We learned in the previous section that VGG8 and ResNet-18 base models pretrained on Birds 25 did perform somewhat better than training a similar model from scratch, producing output comparable to that found by 20× data augmentation. Let's experiment with the new embeddings. Will we see further improvement?

6.3.2   Using the Embeddings with Top-Level Models

The file *imagenet.py* trains top-level models on the embeddings from the previous section. The code is straightforward. Here, I'm summarizing for MobileNet and MLP, the full code is only slightly more involved:

```
xtrn = np.load("results/features_mobilenet/xtrain.npy")
xtst = np.load("results/features_mobilenet/xtest.npy")
ytrn = np.load("results/features_mobilenet/ytrain.npy")
ytst = np.load("results/features_mobilenet/ytest.npy")

clf = MLPClassifier(hidden_layer_sizes=(opt,opt//2), max_iter
    =1000)
clf.fit(xtrn, ytrn)

pred = clf.predict(xtst)
cm,acc = ConfusionMatrix(pred, ytst, num_classes=6)
mcc = matthews_corrcoef(ytst, pred)
print(cm)
print("Test set accuracy: %0.4f, MCC: %0.4f" % (acc,mcc))
```

Load the MobileNet feature vectors, then define the MLP and train. Finally, pass the test set features through the model to calculate the confusion matrix, accuracy and MCC.

I ran *imagenet.py* ten times each for MobileNet and ResNet-50 features training an MLP 512 to find the following mean overall accuracies:

MobileNet	0.78839 ± 0.00232
ResNet-50	0.71438 ± 0.00325

Recall that the *bird6* images were not augmented, meaning the pretrained models produced features leading to considerable improvement over even the Birds 25 pretrained models.

Large models pretrained on ImageNet produce features capturing essential characteristics of the birds. With 59 bird classes among ImageNet's 1000, this isn't too surprising, but is welcome. A brief summary of the best *bird6* results since Chapter 4 seems required:

40 percent   (unaugmented)
↓
57 percent   (augmented 20×)
↓
60 percent   (ensemble, augmented 20×)
↓
79 percent   (ImageNet pretrained, unaugmented)

The final line is marked "unaugmented." Does augmenting *bird6* before producing MobileNet or ResNet-50 embeddings help? Augmentation didn't seem to offer much when working with Birds 25 pretrained features. You can augment the images before extracting features by changing the 1 to a 20 on the *imagenet_features.py* command line. If you do, you'll find a modest accuracy increase of about 3 percentage points.

Careful application of deep learning techniques has effectively doubled model accuracy without requiring new data. A model that is 80 percent accurate is perhaps useful, especially with a properly chosen decision threshold, but we'd like to do better, if possible. Fortunately, we have yet one more trick up our sleeves.

6.4   Using CLIP

In 2021, OpenAI released CLIP (Contrastive Language-Image Pretraining), a model trained on a vast dataset of image and text description pairs. The word "contrastive" refers to the kind of loss used to drive model training. The model includes a vision transformer, an application to computer vision of the same kind of transformer network at the heart of large language models like ChatGPT or Claude. CLIP's joint training on images and associated labels is transformative (pun intended), as the experiments in this section demonstrate.

Accessing CLIP requires installing the PyTorch and CLIP Python libraries:

```
> pip3 install torch
> pip3 install clip
```

CLIP creates embeddings from images and text prompts. It's power comes from the fact that the embedding space is shared, meaning the content of images and the semantics of the text prompt are strongly related. For example, the embeddings produced by images of a Belted Kingfisher and American White Pelican, along with the text prompt "Belted Kingfisher" are, thought of as vectors in a 768-dimensional space, pointing in a particular direction where we expect the angle between the text vector and the kingfisher image vector to be less than the angle the text vector makes with the pelican vector.

The code in *clip_example.py* demonstrates this. First, run the code to produce:

```
Magnitudes:
    text : 11.94
    kingfisher: 17.62
    pelican : 18.84

Cosine distance:
    kingfisher: 0.7233
    pelican : 0.8929

Angles (degrees):
    kingfisher: 73.94
    pelican : 83.85
```

The output displays the magnitudes of the 768-dimensional embedding vectors, the cosine distance between the image vectors and the text vector and the angle between them. The images are read from *kingfisher.png* and *pelican.png*. The magnitude of a vector, also called the norm, is the distance from the origin thinking of the vector as a point in an *n*-dimensional space.

The dot product between two vectors is the sum of the products of the individual vector elements. The magnitude of a vector is the square root of the dot product with itself.

The dot product is also found by multiplying the vector magnitudes along with the cosine of the angle between them. This relationship between vectors holds regardless of the number of dimensions. Therefore, if we have two vectors, *a* and *b*, the cosine of the angle (θ) between the two vectors is:

$$\cos \theta = \frac{a \cdot b}{\|a\|\|b\|}$$

The dot product divided by the product of the vector magnitudes lies within $[-1, 1]$ where vectors pointing in the same direction have a dot product of 1, those at a right angle to each other have a dot product of 0, and those pointing in opposite directions a dot product of -1. Therefore, we can define a *cosine distance* measure as:

$$\text{cosine distance} = 1 - \cos \theta = 1 - \frac{a \cdot b}{\|a\|\|b\|} \tag{6.1}$$

implying that vectors pointing in the same direction have a cosine distance of 0 and those pointing in opposite directions a distance of 2. We'll make extensive use of this metric, and we are now able to fully interpret the output of *clip_example.py* which tells us that the embedding produced by the kingfisher image is closer to the embedding produced by the text "Belted Kingfisher" than the embedding produced by the pelican image as it has a smaller cosine distance, which leads directly to a smaller angle between the two, 74 degrees versus 84 degrees for the pelican image.

Let's review the critical parts of *clip_example.py* before jumping to experiments with the *bird6* dataset. Doing so sets the stage and shows us how to generate CLIP embeddings; consider Listing 6.2.

```
import numpy as np
import torch
import clip
from PIL import Image

device = "cuda" if torch.cuda.is_available() else "cpu"
model, preprocess = clip.load("ViT-L/14@336px", device=device)

text = "Belted Kingfisher"
tokens = clip.tokenize([text]).to(device)
with torch.no_grad():
    text = model.encode_text(tokens).cpu().numpy().squeeze()

image = Image.open("kingfisher.png").convert("RGB")
image = preprocess(image).unsqueeze(0).to(device)
with torch.no_grad():
    king = model.encode_image(image).cpu().numpy().squeeze()

image = Image.open("pelican.png").convert("RGB")
image = preprocess(image).unsqueeze(0).to(device)
with torch.no_grad():
    pelican = model.encode_image(image).cpu().numpy().squeeze()

mag_text = np.linalg.norm(text)
mag_king = np.linalg.norm(king)
mag_pelican = np.linalg.norm(pelican)

d_king = 1.0 - np.dot(king, text) / (mag_king * mag_text)
d_pelican = 1.0 - np.dot(pelican,text)/(mag_pelican*mag_text)

angle_king = np.arccos(1-d_king) * (180/np.pi)
angle_pelican = np.arccos(1-d_pelican) * (180/np.pi)
```

Listing 6.2 Configuring and using CLIP

The imports come first. NumPy, as usual, followed by PyTorch, then CLIP, and finally the Image module from Pillow. CLIP works with Image objects, not NumPy arrays.

The following two lines configure PyTorch and CLIP. No GPU is necessary. The latest CLIP model is loaded and downloaded on the first run.

We need an embedding of the text "Belted Kingfisher," hence the next four lines of code, first to define the text, then to break it into tokens before overwriting text with the 768-element embedding vector returned by CLIP. The with statement creates a context manager to use PyTorch without wasting effort calculating gradients that might be used for updating the model weights.

We also need embeddings for the two images we want to compare. The images are read from disk as Image objects, then passed through CLIP using the encode_image method on the model object.

The necessary embedding vectors are now in place, one for the text string "Belted Kingfisher" and two for the images. The vectors are roughly symmetric, about zero, with means near zero. However, a plot of the vectors indicates that while most elements are within [−1, 1] or so, there are several "spikes" where elements have larger values, though all in this case lie within [− 10, 10]. Somehow, all of these numbers encode a significant amount of information about the images and text string, information related to context and meaning. CLIP's contrastive loss training has also forced the model to develop meaningful directions in the embedding space so that similar concepts are directed in similar directions. And that fact allows us to use the embeddings to build a simple classifier.

The embeddings exist in a 768-dimensional space, meaning we need 768 numbers in each embedding. We might think of these numbers as the coordinates of a point in the 768-dimensional space. Alternatively, we might think of them as specifying a vector, an arrow in 768 dimensions, beginning at the origin, (0, 0, 0, …, 0), and ending at the point indicated by the embedding vector values.

If CLIP training has aligned embedding vectors along some abstract set of concepts indicated by each dimension, or more likely mixed among the dimensions, then we might expect embedding vectors of related concepts to point in much the same direction. Therefore, one approach to labeling images cast as embedding vectors is to assign them to the text corresponding to the text embedding that most closely aligns with the image embedding. In other words, pick the label of the embedding pointing in much the same direction as the image embedding. All of which is to say that we select the label of the embedding with the smallest cosine distance for each unknown image embedding. Recall that a small cosine distance implies a small angle between the two vectors.

The final part of Listing 6.2 calculates the cosine distances and angles between the image embeddings and the text embedding,

```
mag_text = np.linalg.norm(text)
mag_king = np.linalg.norm(king)
mag_pelican = np.linalg.norm(pelican)

d_king = 1.0 - np.dot(king, text) / (mag_king * mag_text)
d_pelican = 1.0 - np.dot(pelican,text)/(mag_pelican*mag_text)

angle_king = np.arccos(1-d_king) * (180/np.pi)
angle_pelican = np.arccos(1-d_pelican) * (180/np.pi)
```

First, calculate the length of each vector, its magnitude (or norm). We'll use NumPy's norm function for convenience. The cosine distances come next following Equation 6.1. The corresponding angles, in degrees, are then found by solving for θ, leading to the previously given output:

```
Cosine distance:
    kingfisher: 0.7233
    pelican : 0.8929

Angles (degrees):
    kingfisher: 73.94
    pelican : 83.85
```

The cosine distance between "Belted Kingfisher" and the kingfisher image is less than the distance to the pelican image embedding, therefore, a simple classifier looking for the nearest text label would, correctly, assign the kingfisher image to the Belted Kingfisher class. The angles indicate the same.

We'll use this approach in the experiments of this section, but first, let's appreciate what the numbers tell us. CLIP, trained on a massive collection of images and associated text, has captured, via its model architecture driven by the contrastive loss, sufficient information about bird-ness, visual and linguistic, to associate the two very different inputs. Therefore, we can accurately pair the image with a known text label without further refinement. Just what the embedding has captured is beyond complete understanding at present, though clever explorations of LLMs have uncovered rough collections of concepts learned by models during training. The concepts paint a picture of a dense representation, where concepts are spread over different embedding values in a kind of "superposition," to borrow a term from quantum physics.

Exploring CLIP embeddings implies having CLIP embeddings. Therefore, we begin by generating embedding vectors for the *bird6* dataset. After that, we use the embeddings to train a top-level MLP, allowing for a direct comparison with the ImageNet pretrained embeddings in the previous section.

Subsequent experiments return to the notion of comparing cosine distances, first with a few "template" images and then with species names and descriptions in text. We end with an experiment attempting a more fine-grained classification of visually similar species.

6.4.1 CLIP Embeddings

Creating *bird6* embeddings (features) is straightforward: for each train and test image, resize to 336×336 pixels, pass the image through the CLIP model and accumulate the resulting collection of embeddings as feature vectors. The file *clip_features.py* contains the necessary code with the CLIP function the essential portion:

```
def CLIP(images, model, preprocess):
    ans = []
    for x in images:
        img = Image.fromarray(x).convert("RGB")
        img = img.resize((336,336), resample=Image.BILINEAR)
        im = preprocess(img).unsqueeze(0).to(device)
        with torch.no_grad():
            features = model.encode_image(im)
        ans.append(features.cpu().numpy().squeeze())
    return np.array(ans)
```

The *bird6* images as passed, along with the configured CLIP model and preprocess routine. Each image is mapped from NumPy to Pillow (Image), then resized from 64×64 to 336×336 to best align with what CLIP expects. The resulting collection of embedding vectors is then returned. The main code in *clip_features.py* generates embeddings and stores them, and the associated class labels, in the given output directory. For example:

```
> python3 clip_features.py results/features_clip
```

creates feature files in the directory subsequent experiment code expects:

```
bird6_clip_xtest.npy
bird6_clip_xtrain.npy
bird6_clip_ytest.npy
bird6_clip_ytrain.npy
```

The *bird6* images are now cast as embedding vectors and we are ready for some experiments.

6.4.2 CLIP Classifiers

Let's compare the CLIP embeddings to the ImageNet pretrained embeddings from earlier in the chapter by training a top-level MLP of the same size. The necessary code is in *clip_classifier.py*, which is brief and easy to follow.

For example, one run produced:

```
> python3 clip_classifier.py MLP 512

MLP 512, CLIP features:
[[ 82   0   2   0   1   0]
 [  0  29   0   0   0   0]
 [  0   0  23   0   0   0]
 [  0   0   3  77   0   0]
 [  4   1   1   0 127   0]
 [  0   0   0   3   0  43]]
Test set accuracy: 0.9621, MCC: 0.9518
```

CLIP embeddings led to a dramatic increase in overall accuracy on the *bird6* test set. Training ten such models returns the following mean accuracy including the previous ImageNet results for comparison:

CLIP	0.94291 ± 0.00196
MobileNet	0.78839 ± 0.00232
ResNet-50	0.71438 ± 0.00325

Recall that these results are for the unaugmented version of the *bird6* dataset and represent yet another significant jump in average performance of over 15 percentage points. Such a model is clearly fit to task and is a testament to the information capacity of CLIP features.

The code in *clip_classifier.py* uses embeddings to train a classical machine learning model. Let's shift gears to focus instead on cosine distance to develop even simpler models.

Using Image Templates

The fact that the cosine distance between an embedding of the phrase "Belted Kingfisher" and the embedding of an image of a kingfisher was less than the distance to the embedding of a pelican points the way toward a simple model: assign the label of the closest label embedding. Essentially, this is a *k-nearest neighbor (NN)* model with $k = 1$.

Let's proceed as follows. First, we'll select a set of bird images to use as templates, three for each of the six classes in *bird6*. I chose public domain images from Wikimedia Commons, running them through *chipper.py* to output 336×336 pixel images. The image embeddings and corresponding class labels form a reference dataset over which we can search for the closest match (smallest cosine distance) for an unknown bird image embedding. The code in *clip_template_features.py* processes the template images and stores the embeddings and labels on disk.

Classification becomes: for each *bird6* test set embedding, calculate the cosine distance to each of the template embeddings and assign the label of the closest embedding. We know the true labels and can construct a confusion matrix and associated metrics. The necessary code is in *clip_template_classifier.py*, of which the critical function is:

```
def Classify(x,y, xtst, ytst):
    pred = []
    for i in range(len(xtst)):
        mn, lbl = 2,0
        for j in range(len(x)):
            sc = Cosine(x[j], xtst[i])
            if (sc < mn):
                mn,lbl = sc, y[j]
        pred.append(lbl)
    pred = np.array(pred)
    cm, acc = ConfusionMatrix(pred, ytst, num_classes=6)
    mcc = matthews_corrcoef(ytst, pred)
    return cm, acc, mcc
```

where the template embeddings and labels are in x and y. The double `for` loop is a direct implementation of the algorithm assigning the label of the closest template embedding.

The code performs three passes, beginning by using all three template embeddings per class individually.

The second pass uses a single template image, randomly the first in the set of three, selecting among six templates, one per class.

The final pass first averages the three templates per class, element by element, and then uses those six mean templates as references before outputting the results for all three passes.

A sample run produced the following with three templates per class:

```
3 templates per class:
[[ 67   0   2   0  16   0]
 [  0  29   0   0   0   0]
 [  0   0  22   0   1   0]
 [  0  18   0  59   2   1]
 [  8   0   0   3 122   0]
 [  0   1   0   4   0  41]]
Test set accuracy: 0.8586, MCC: 0.8215
```

An overall accuracy of about 86 percent isn't terrible, but notice that certain classes are often confused, namely kestrels (class 0) with red-tails (class 4) and blue herons (class 3) with pelicans (class 1).

The single template results are similar overall:

```
1 template per class:
[[ 54   0   3   2  26   0]
 [  0  23   0   2   0   4]
```

```
[  0    0   22    0    1    0]
[  0    0    2   73    5    0]
[  0    0    0    4  125    4]
[  0    0    0    0    0   46]]
Test set accuracy: 0.8662, MCC: 0.8325
```

but notice that the confusion between herons and pelicans has disappeared. At first, it is counterintuitive why a single template per class works better than three templates. One possible explanation is that the chipped template images might not capture enough of the larger birds, especially if capturing the head of the herons and pelicans leaves the long bill without sufficient context to distinguish between the two. In this case, the single template image for the pelican includes all of the bird while the heron focuses on the head, likely enough of a distinction to eliminate the confusion.

There is no a priori reason to believe that the template images are in any way the best representatives of which direction in embedding space most images of each species point. Therefore, it seems reasonable to average the three per-species embedding vectors and use those to select a class label. This is precisely what the final output from the template classifier represents:

```
Mean template class:
[[ 78    0    0    0    7    0]
 [  0   29    0    0    0    0]
 [  0    0   22    0    1    0]
 [  0    3    0   73    4    0]
 [  1    0    0    0  132    0]
 [  0    0    0    4    0   42]]
Test set accuracy: 0.9495, MCC: 0.9355
```

An overall accuracy of nearly 95 percent is validation of that intuition behind averaging the template image embeddings given the nearly ten percentage point jump in overall accuracy.

Working with template images makes sense, but CLIP is a multimodal model, so why restrict ourselves to template images when we can use "template" label text? Let's give that thought a try.

Using Text Embeddings

Let's replace the template images with a single text label for each class. The obvious choice is the bird's species name, here the common name. The code in *clip_text_classifier.py* gives us what we need. Execution produces three output confusion matrices, the first is:

```
Common name:
[[ 83    0    0    0    2    0]
 [  0   29    0    0    0    0]
 [  0    0   23    0    0    0]
 [  3    0    6   71    0    0]
 [  4    0    0    0  129    0]
 [  0    0    0    0    0   46]]
Test set accuracy: 0.9621, MCC: 0.9521
```

proving that comparing bird image embeddings to common name embeddings is definitely a worthwhile exercise producing an overall accuracy of 96.2 percent on the *bird6* test set; a far cry from the 40 percent accuracy we first found in Chapter 4.

It's likely that CLIP was trained on bird images using the common name. If so, we might expect CLIP to be just as effective with the scientific name. The second confusion matrix tests this assumption:

```
Scientific name:
[[ 43    0    0    0   42    0]
 [  0   29    0    0    0    0]
 [  4   11    0    6    2    0]
 [  1   11    0   62    4    2]
 [  0    0    0    0  133    0]
 [  0    2    0    2    0   42]]
Test set accuracy: 0.7803, MCC: 0.7272
```

The results are curious. Red-tailed Hawks (class 4) are perfectly detected, but Belted Kingfishers (class 2) are never correctly classified and the kestrels are back to being confused with red-tails. I take these results as an indication that there are many more images in CLIP's training set paired with common names instead of scientific ones.

CLIP text embeddings may be up to 77 tokens in length or about 50 words. I asked OpenAI's GPT-4o to describe each *bird6* species briefly. For example, here's how it described red-tails:

> A robust raptor with a pale, streaked underside and a warm, reddish-brown tail that contrasts sharply with its dark brown back and wings. The head is broad and dark with a hooked yellow-tipped bill, while its legs are feathered only partway down, ending in yellow talons.

The descriptions are within the source code. Their embeddings led to the following:

```
Text description:
[[ 66    0    0    0   19    0]
 [  0   11    0    1    0   17]
 [  5    1   15    1    0    1]
 [  2    0    3   42   21   12]
 [  5    0    0    0  128    0]
 [  0    0    0    0    0   46]]
Test set accuracy: 0.7778, MCC: 0.7215
```

Let's consider these results carefully. Intuitively, we expect some level of performance using template images, which is essentially in the same vein as building a training set, and similarly with text labels based on the common name. However, achieving nearly 78 percent accuracy when simply describing the bird without using its name uncovers some of the power hidden in CLIP embeddings. It's not merely a shared space mapping images and text to a set of numbers, but a space where *meaning* has emerged so that what describes the bird is also aligned with images of the bird.

The *bird6* dataset contains birds that are easy to distinguish visually. Are CLIP features useful for the harder task of fine-grained classification among highly similar species? Let's find out.

Fine-Grained Classification

CLIP embeddings performed well when classifying visually distinct bird species. Let's put them to the test with sparrows. The file *sparrow_features.npy* is the output of *clip_sparrow_features.py*, which processes a small image dataset of nine sparrow species; see Figure 6.3.

Classification with these features relies on the species' common name as found in *clip_sparrow_classifier.py*. The output is a confusion matrix:

```
Sparrows by common name:
[[2 0 0 0 0 0 1 0 0]
 [1 4 0 0 0 0 0 0 0]
 [1 0 3 0 0 0 0 0 0]
 [0 0 0 9 0 0 0 0 0]
 [0 0 0 0 1 0 0 0 0]
 [0 0 0 1 0 2 0 0 0]
 [0 0 0 0 0 0 5 0 0]
 [0 0 0 0 6 0 1 7 0]
 [0 0 0 0 0 0 0 0 7]]
Test set accuracy: 0.7843, MCC: 0.7691
```

The class labels match those in Figure 6.3.

An accuracy of over 78 percent is a pleasant surprise. Most of the errors come from confusing Vesper Sparrows (class 7) with Lincoln's Sparrows (class 4). Other sparrow species are more likely to be correctly labeled. Of course, the dataset is small, but it is nonetheless successful in demonstrating how CLIP embeddings are useful for tasks that humans find difficult.

6.5 Discussion

We ignored fine-tuning in this chapter for a reason: CLIP embeddings are exceptionally effective for bird classification, no fine-tuning is required. CLIP isn't the only game in town; there are many image-text multimodal models out there, for example OpenCLIP or LAVIS from Salesforce.

It is possible to fine-tune CLIP-like models via techniques like LoRA (*low-rank adaptation*), but those techniques are far beyond what we should consider here. We'll continue to explore CLIP embeddings throughout the remainder of the book, but know that other models are available and worth experimentation, should you begin a research project benefiting from such embeddings.

Two of the chapter's experiments produced essentially identical results, both achieving 96 percent test set accuracy on *bird6*. The first trained a top-level MLP using *bird6* training set CLIP embeddings, while the second relied on nearest-neighbor classification using the cosine distance between common name embeddings and the image embeddings. The two approaches produced effective models, but the former required a labeled training set. Building a training set is labor-intensive, as we discovered in Chapter 4.

Figure 6.3 Left to right, top to bottom: Brewer's (0), Chipping (1), House (2), Lark (3), Lincoln's (4), Savannah (5), Song (6), Vesper (7), White-crowned (8)

The common name classifier required no training data whatsoever. Pretrained CLIP alone was sufficient to produce information-rich embeddings that could correctly label the diverse species of *bird6* and effectively separate visually similar sparrow species. As mentioned earlier, it is highly likely that CLIP was trained on many bird image and label pairs. The relatively poor performance found when using the scientific name supports this view and points to a situation where LoRA-style fine-tuning of CLIP, and other models might be effective should a system operating with scientific names be the goal.

The common name experiment used CLIP as a *zero-shot* classifier. Zero-shot refers to applying an unaltered model to tasks for which it was not explicitly trained. The CLIP training session unintentionally already conditioned the model to use cases like identifying birds in images. Think of it as a kind of emergent ability from the more general task of aligning images and text of all kinds.

The arrival of models like CLIP has opened new vistas by greatly simplifying what was previously a sometimes difficult deep learning task: gathering sufficient training data to successfully develop a CNN or similar model from scratch. We'll continue to take advantage of CLIP's power in the next chapter's experiments where we examine the feasibility of general-purpose bird classifiers.

7. Generic Bird Classifiers

The previous chapter demonstrated the utility of CLIP features. In this chapter, we go for broke and attempt to develop what might be called "generic bird classifiers." We'll be partially, even mostly, successful, but will also uncover weaknesses and explore approaches that might lead to better results with more effort. And, as always, we'll increase our expertise with modern AI along the way.

The heart of the chapter involves a set of CLIP image features extracted from the *NABirds* dataset (https://dl.allaboutbirds.org/nabirds). The dataset contains some 48,000 photographs of over 400 North American bird species (404, to be exact). We won't use all 48,000 images; in fact, a mere 16 per class will prove sufficient.

With *NABirds* features in hand, we're ready for classification experiments. We continue as in Chapter 6 with a combination of classifiers relying on the cosine distance between image and text embeddings and MLP and support vector machines trained on the CLIP embeddings.

From these experiments, we finally arrive at a set of "generic" models capable of classifying arbitrary bird images and text descriptions. The chapter ends with a discussion to put everything into context.

7.1 North American Bird Features

We know that CLIP embeddings perform brilliantly when used to classify the *bird6* images. However, the six species in that dataset are highly distinct. The sparrows exercise in Chapter 6 hints that CLIP embeddings might be applicable to differentiating similar species. Building on that foundation requires a new dataset involving many more species. Fortunately, such a dataset exists: the North American birds dataset found at https://dl.allaboutbirds.org/nabirds.

As we'll refer to it, the *NABirds* dataset consists of over 48,000 images of 404 species of North American birds separated by sub-categories related to visual appearance. The downloaded dataset occupies 9.5 GB of disk space when fully expanded.

Let's use the images to build a collection of CLIP features for the chapter's experiments. Thankfully, you do not need to download the dataset: I already did and used it with the code described in this section to build the necessary CLIP files:

```
nabirds_features.npy
nabirds_labels.npy
nabirds_names.npy
```

The files contain 8877 CLIP features, their corresponding class labels and the common names of the 404 species in the dataset. The feature file is distributed with permission and in accordance with the *NABirds1 Terms of Use*, which requires the following notice:

> Data provided by the Cornell Lab of Ornithology, with thanks to photographers and contributors of crowdsourced data at AllAboutBirds.org/Labs. This material is based upon work supported by the National Science Foundation under Grant No. 1010818.

If you wish, jump to the next section to begin the experiments. Otherwise, read on to peruse the nitty-gritty behind the features.

After expanding the dataset, I explored how it was presented on disk, then wrote code to gather 16 instances of each of the images stored in the 716 directories corresponding to the different visual classes. Visual classes separate between, say, juvenile and adult birds of the same species. I bundled these together as instances of the species to pare the 716 categories down to 404 classes; see Listing 7.1.

```
from random import shuffle, seed
seed(8675309)
classes = [
    ( 295, "Common Eider"),
    ### 402 additional classes ###
    (1010, "Dark-eyed Junco"),
]
names = sorted(list(set([i[1] for i in classes])))

N = 16
xnames, ylabels = [], []
for d,name in classes:
    if (name in names):
        yl = names.index(name)
    else:
        raise ValueError("unknown: %s" % name)
    base = "nabirds/images/%04d/" % d
    try:
        bnames = sorted(os.listdir(base))
        shuffle(bnames)
        bnames = bnames[:N]
        for n in bnames:
            xnames.append("%s%s" % (base,n))
            ylabels.append(yl)
    except:
        pass

ylabels = np.array(ylabels)
names = np.array(names)
np.save("nabirds_labels.npy", ylabels)
np.save("nabirds_names.npy", names)
```

Listing 7.1 Extracting image names

Listing 7.1 is split into three code paragraphs. The first is a long list of directory names (as numbers) and the associated common name. The directories contain bird images for that species. The variable names contains a sorted list of the 404 names.

The second code paragraph walks through the class list and extracts the pathname of 16 randomly selected images. The seed is fixed to produce the same set on each run. Notice mild error-checking code to capture any inconsistencies in how the data is stored. Such often happens in practice, though the dataset builders were thorough in this case, and no inconsistencies were present.

The third paragraph turns the collection of pathnames and labels (indices into the sorted name list) into NumPy arrays and stores them on disk.

The variable xnames contains pathnames to specific bird images. The *NABirds* dataset uses bounding boxes to outline the bird in each image. We need this information to chip out the bird.

Listing 7.2 walks through xnames to load the full image file (im) then use the associated bounding box information to extract the bird portion of the image (img). Ultimately, a set of 8877 bird images is stored on disk. CLIP embeddings are generated from these chips by *nabirds_features.py*, which follows the same process used for *bird6* embeddings.

```
t= [i[:-1].split() for i in open("nabirds/bounding_boxes.txt")]
boxes = {}
for s,x,y,w,h in t:
    s = s.replace("-","")
    b = (int(x),int(y),int(w),int(h))
    boxes[s] = b

x = np.zeros((len(xnames),336,336,3), dtype="uint8")
for k in range(len(xnames)):
    im = np.array(Image.open(xnames[k]).convert("RGB"))
    bn = os.path.basename(xnames[k])[:-4]
    bb = boxes[bn]
    im = im[bb[1]:(bb[1]+bb[3]),bb[0]:(bb[0]+bb[2]),:]
    w,h,_ = im.shape
    if (w > h):
        img = np.zeros((w,w,3), dtype="uint8")
        offset = (w-h) // 2
        img[:,offset:(offset+h),:] = im
    else:
        img = np.zeros((h,h,3), dtype="uint8")
        offset = (h-w)// 2
        img[offset:(offset+w),:,:] = im
    img = Image.fromarray(img).resize((336,336),
                resample=Image.BILINEAR)
    x[k] = np.array(img)
np.save("nabirds_images.npy", x)
```

Listing 7.2 Extracting images with bounding boxes

We intend to classify bird images by comparing their CLIP embedding to the embedding of the bird's common name. Therefore, we need a set of name embeddings; see *nabirds_name_embeddings.py*.

Also, we want to explore classifying CLIP embeddings by using the average embedding produced by the image features for a class. For example, we want the average CLIP embedding representing all the Great Blue Heron image features. The code in *nabirds_average_features.py* does this for us by loading the images and common name features before averaging across classes, first the image features alone, then again including the common name feature as well under the belief that averaging the text name embedding with the image embeddings might improve classification accuracy.

We have image and name features. Let's apply them to learn whether simple classifiers can operate as generic bird image classifiers.

7.2   Using NA Bird Features

The previous section left us with 768-element feature vectors in *nabirds_features.npy* and their associated labels in *nabirds_labels.npy* along with a set of 404 class names (*nabirds_names.npy*) and their text embeddings (*nabird_name_embeddings.npy*). These are all we need to begin building classifiers.

The file *nabirds_classifier.py* implements a nearest-neighbor cosine distance model. It works with *bird6* and *NABirds* image embeddings; ignore the large source option for now. I'll introduce it later in the chapter.

Two additional classifiers feature in this section: one using an MLP (*nabirds_mlp_classifier.py* and another using a support vector machine (*nabirds_svm_classifier.py*). Both split the *NABirds* features into train and test sets to create the top-level model. We'll explore all three before diving deeper into the results and code.

7.2.1   Running the *NABirds* Model

The code in *nabirds_classifier.py* expects a data source and an output file directory. For example, to run against *bird6*, use a command line like:

```
> python3 nabirds_classifier.py bird6 classifier_bird6
```

which compares the 404 class *NABirds* class names, as embeddings, to each of the *bird6* test set image embeddings. You should see the following output as the model is deterministic:

```
Common name (bird6):
[[0 0 0 ... 0 0 0]
 [0 0 0 ... 0 0 0]
 [0 0 0 ... 0 0 0]
 ...
 [0 0 0 ... 0 0 0]
 [0 0 0 ... 0 0 0]
 [0 0 0 ... 0 0 0]]
Test set accuracy: 0.4470, MCC: 0.4449
```

```
American Kestrel (0.776):
American Kestrel      (66)
Sharp-shinned Hawk   ( 6)
Prairie Falcon       ( 4)
Broad-winged Hawk    ( 2)
Say's Phoebe         ( 2)
```

Note that I've adjusted the format to fit the page. You will see a single line of output for each of the six classes after the confusion matrix, not merely the single example above. Notice that the overall test set accuracy is 44.7 percent, a far cry from the 96.2 percent a similar approach returned in Chapter 6.

Each output line begins with the name of a *bird6* class followed by the class-level accuracy. For the kestrel example, the per-class accuracy was 77.6 percent.

What follows are the first five most often selected class names. In other words, the first five classes to which American Kestrel images (feature vectors) were most often assigned. The example indicates that the cosine distance model has correctly identified 66 of the 85 kestrel images in the test set and that kestrels were most often confused with Sharp-shinned Hawks (6 times) followed by Prairie Falcons (4 times). Notice that neither species is part of the *bird6* dataset.

The remaining *bird6* classes follow suit with per-class accuracies and most often assigned class of:

```
American White Pelican  (0.931): American White Pelic  ( 27)
Belted Kingfisher       (0.609): Belted Kingfisher     ( 14)
Great Blue Heron        (0.375): Great Blue Heron      ( 30)
Red-tailed Hawk         (0.023): Broad-winged Hawk     ( 59)
Snowy Egret             (0.804): Snowy Egret           ( 37)
```

Each class, with one exception, is most often correctly assigned. Note that this doesn't imply high accuracy, as there are over 400 class labels to which the test sample might be assigned. Pelicans achieved a 93 percent per-class accuracy, while herons, most frequently called herons, managed only a 37.5 percent accuracy. Red-tails fail miserably, with only 2.3 percent of the 133 in the test set correctly labeled.

Red-tails are most often categorized as Broad-winged Hawks (59 times) followed by Swainson's Hawks (31 times) and Rough-legged Hawks (22 times). Subsequent errors are in the single digits per class. Why red-tails fail will be explored in the next section. Let's continue with the files *nabirds_classifier.py* dumped in the output directory:

```
confusion_matrix.npy
console.txt
correct.npy
distances.npy
labels.npy
per_class.txt
predictions.npy
```

Some of the output is familiar, like the 404×404 confusion matrix, the per test set sample predictions and true label, along with the entire console output. Three of the files are new. The file *correct.npy* holds a vector where elements are 1 if the corresponding test sample was correctly labeled and 0 otherwise.

Look at the vector in *distances.npy* to find the cosine distance between that sample's image embedding and the common name embedding of the assigned label.

Finally, the file *per_class.txt* contains the text output for each class along with the five most commonly assigned labels. In other words, the file contains the text generated after the confusion matrix and overall accuracy.

There are only 6 classes in *bird6* but 404 in *NABirds*. Will the nearest-neighbor cosine distance classifier work with so many classes? Recall that we are using CLIP out of the box, therefore, all the *NABirds* image features are available to us as test data:

```
> python3 nabirds_classifier.py nabirds classifier_nabirds
```

The output includes the overall accuracy and MCC:

```
Test set accuracy: 0.6061, MCC: 0.6053
```

plus output text for each of the 404 classes.

An overall accuracy of 60.6 percent isn't at first take impressive until we remember that there are 404 classes implying a random guess accuracy of about 0.25 percent on average. Therefore, CLIP features and cosine distance are performing rather well.

7.2.2   Training Top-Level Models

We can use CLIP features as training data for top-level models, provided we split the available data into train and test. Each *NABirds* class has at least 16 samples by design. Some *NABirds* classes are variants of the more basic common name, meaning a few of the 404 classes have more than 16 samples. Let's reserve 10 samples of each class for training and use the remaining 6 as test. As percentages, this means 62.5 percent train and 37.5 percent test. Splitting happens on the fly when the MLP and SVM classifiers are created, though the pseudorandom number seed is fixed to repeat the split on each run.

The code in *nabirds_mlp_classifier.py* and its SVM partner produce the same output as *nabirds_classifier.py*, all that changes is how the output is determined. The command lines are the similar as well, with the MLP version accepting the number of nodes in the first hidden layer before the data source. For example, to train top-level models and apply them to *bird6* run command lines like:

```
> python3 nabirds_mlp_classifier.py 512 bird6 mlp_512_bird6
> python3 nabirds_svm_classifier.py bird6 svm_bird6
```

The first trains an MLP with 512 nodes using *NABirds* features, then applies the model to the *bird6* test data. The 512 nodes of the first hidden layer match well with the 768 elements of the input feature vector. The second repeats the process with a support vector machine.

The output directories have most of the files produced by the cosine distance model, except *distances.npy*. Additionally, the trained MLP or SVM is stored as a Python `pickle` file (*.pkl*) for future use. Table 7.1 contains a summary of the results.

All three models perform (seemingly) poorly on *bird6*. To understand why, read on to the next section. The full 404 class *NABirds* models fare better. Note that the nearest-neighbor

Table 7.1 The overall accuracy of the CLIP embedding models.

Model	bird6	NABirds
Cosine distance	0.4470	0.6061
MLP 512	0.5152	0.8081
SVM	0.4040	0.8189

cosine distance model model assessed all 8877 *NABirds* feature vectors while the MLP and SVM models trained on 62.5 percent of that data and tested the remaining 37.5 percent. Still, random sampling in splitting the *NABirds* features imply that there should be no meaningful difference, statistically speaking, between the train and test features.

CLIP features with the simple nearest-neighbor cosine distance classifier produce an impressive 60.6 percent accuracy over 404 classes. ImageNet-based models with 1000 classes didn't exceed that percentage of accuracy until 2012 with the advent of CNNs (AlexNet) specially designed for that purpose. CLIP was trained on generic image and text pairs and is not tailored to any class of problem in particular.

The MLP and SVM results boost performance to just shy of 82 percent over the same 404 classes. Recall that most classes were represented in the training set by a mere 10 samples, yet the CLIP embeddings are sufficiently information-rich to allow simple top-level models to effectively learn to recognize bird species from only a handful of examples. Human infants similarly learn to recognize objects from only a few features. Might they also be mentally building something akin to CLIP features in their minds, especially during the first year of life? A provocative thought, and not my own: see *Helpless infants are learning a foundation model*, an opinion piece by Rhodri Cusack, Marc'Aurelio Ranzato and Christine J. Charvet (Trends in Cognitive Sciences, Volume 28, Issue 8, August 2024).

The SVM *bird6* model is the worst performer on that dataset. I used a linear kernel SVM and a default C constant of 1. It's possible that tweaking both the kernel and their associated constants, C for linear or C and γ (gamma) for a Gaussian kernel (also known as a radial basis function kernel) might improve matters somewhat, but I expect no better than the cosine distance level of performance.

Overall accuracy is a handy measure if the models are trained and tested on the same data, as these are, but deeper insights await.

7.2.3   Diving Into the Results

The *NABirds* classifiers generate multiple output files. The code in *nabirds_classifier_results. py* analyzes those files to produce additional insights into the results. The only command line argument is the name of a directory containing classifier output. For example:

```
> python3 nabirds_classifier_results.py classifier_nabirds
```

interprets the *NABirds* cosine distance results from the previous section.

Two plots are displayed sequentially. Click the close box in the upper right to move to the next. Text is also generated:

```
Distance, correct vs incorrect:
    correct  : 0.704555 +/- 0.000222
    incorrect: 0.712279 +/- 0.000274

(t= 21.85025854, p=0.00000000, u=0.00000000, Cohen's d=0.475)
```

```
Accuracy:
     overall        : 0.60606
     mean per class: 0.61357 +/- 0.30785 (SD)
     median         : 0.68750 +/- 0.25000 (MAD)
```

Let's begin with the text. The directory contains output from the cosine classifier, implying the existence of a *distances.npy* file. This file, in combination with *correct.npy*, leads to a the mean ± SE cosine distance for correctly and incorrectly classified *NABirds* test samples. The following line presents the t-test and Mann-Whitney U test results where p and u are the associated p-values, respectively. Both are essentially zero, indicating a highly significant difference in the cosine distances between correct and incorrect classifications.

Cohen's *d* measures *effect size*, a quantization of the difference between two means scaled by their pooled standard deviation. A *d* of 0.475 is considered a moderate effect, implying a meaningful difference in means between correctly and incorrectly classified samples.

Accuracy comes next. The overall accuracy is 0.60606 matching Table 7.1. The following two accuracy measurements are new. The first is the mean per-class accuracy over the 404 classes along with its standard deviation (not standard error). The second is the median per class accuracy and the mean absolute deviation (MAD), a measure of variation about the median analogous to the standard deviation's measure of variation about the mean. The relative magnitude of the three accuracy measures is telling. That the overall accuracy is lower compared to the others indicates one or more classes are performing poorly relative to the others. The fact that the mean and median accuracies are quite different hints that per-class accuracies are not symmetrically distributed.

Means are strongly influenced by outliers, while medians are less so. The same is true for their respective measures of variation. The ordering of mean and median also indicates that one or more classes are performing poorly, and half of the classes achieved an accuracy above 68.8 percent. Recall that the median is also the 50-th percentile, implying that some 200 species are reasonably well categorized by the nearest-neighbor cosine distance.

Figure 7.1 contains the two plots produced when evaluating the *NABirds* results. On the top is the distribution of per-class accuracies. The distribution is not symmetric, implying less weight should be given to the mean and more to the median. Further, the distribution is skewed toward higher accuracies, indicating again that many (most) species are well-detected.

The lower plot contains the distribution of cosine distances segregated by correct (red) or incorrect (blue) class assignment. The distributions are normal-like and demonstrate that correctly classified samples involve smaller cosine distances, in general.

Now, re-run *nabirds_classifier_results.py* passing it *classifier_bird6*. The output text is:

```
Distance, correct vs incorrect:
     correct   : 0.716745 +/- 0.001524
     incorrect: 0.727352 +/- 0.000913

(t= 6.22280419, p=0.00000000, u=0.00000000, Cohen's d=0.629)

Accuracy:
     overall        : 0.44697
     mean per class: 0.58635 +/- 0.33631 (SD)
     median         : 0.69258 +/- 0.17511 (MAD)
```

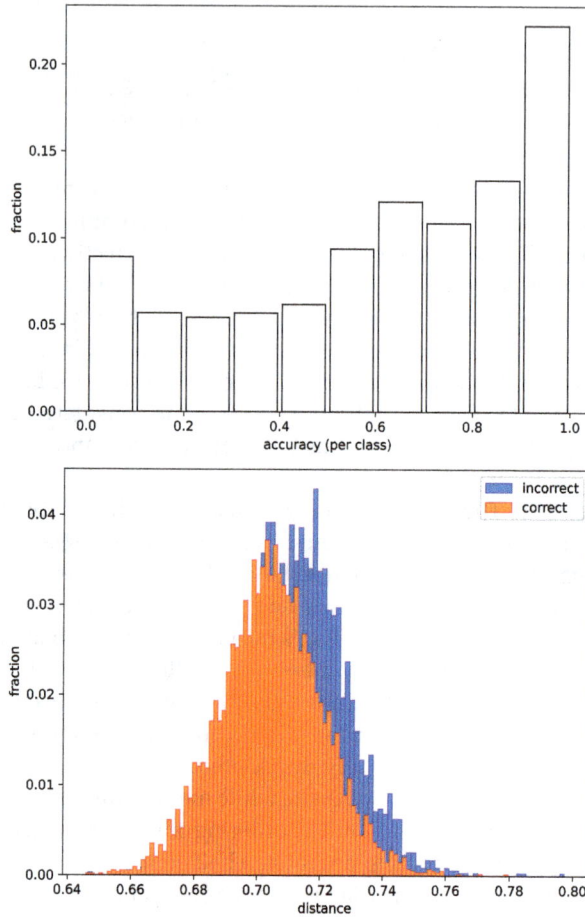

Figure 7.1 Distribution of per-class accuracies (top) and cosine distances by correct or incorrect for *NABirds* (bottom).

There are only six classes and about 400 individual test samples, but the previous statements about accuracy still holds, indicating that half of the classes, in other words, 3 of the 6, achieved per-class accuracies of above 70 percent: American Kestrel, American White Pelican and Snowy Egret. The mean is pulled significantly downward by the Red-tailed Hawk's poor showing with an accuracy of only 2.3 percent.

The distance distribution plot (not shown) is noisy compared to Figure 7.1, but qualitatively similar in that correctly classified samples have, in general, smaller cosine distances than those incorrectly classified. A Cohen's *d* of 0.629 indicates a similar moderate effect size.

The code in *nabirds_classifier_results.py* works with the output of the MLP and SVM classifiers by skipping the cosine distance results and plot. For example, the MLP 512 results for *NABirds* leads to:

```
Accuracy:
    overall        : 0.80805
    mean per class: 0.80559 +/- 0.19405 (SD)
    median         : 0.83333 +/- 0.16667 (MAD)
```

with the SVM results producing:

```
Accuracy:
    overall        : 0.81886
    mean per class: 0.81762 +/- 0.18741 (SD)
    median         : 0.83333 +/- 0.16667 (MAD)
```

The output, and associated per-class accuracy distribution plots (not shown), demonstrate that training a top-level model with even a handful of examples, about 10 per class on average, leads to still greater performance, with 50 percent of the classes exceeding 83.3 percent accuracy. The smaller gap between the SVM mean and median points to a slight edge over the MLP, though remember that the MLP's training is stochastic while the SVM is deterministic in this case. Re-training the MLP might alter the results.

The encouraging *NABirds* results for simple top-level models are not echoed by *bird6* where we find (MLP first, SVM second):

```
Accuracy:
    overall        : 0.51515
    mean per class: 0.52546 +/- 0.23000 (SD)
    median         : 0.49876 +/- 0.17283 (MAD)

Accuracy:
    overall        : 0.40404
    mean per class: 0.46442 +/- 0.23400 (SD)
    median         : 0.40625 +/- 0.09170 (MAD)
```

Here, the top-level models not only did not improve on the cosine distance results but were considerably worse, though the reason isn't immediately evident.

We'll return to these results later in the chapter, but for now, let's briefly investigate the respective classifier code.

7.2.4   Understanding the Code

It's helpful to understand the code behind the three classifiers of this section: nearest-neighbor cosine distance, MLP and SVM. However, a detailed code walkthrough is needlessly pedantic given that we've explored similar classifiers in Chapter 6. Instead, I'll merely point out critical parts of the code for this application and trust your natural curiosity to take care of the rest.

First, we have CLIP embeddings for the 404 classes in *NABirds* where class labels are integers in [0, 403] corresponding to the sorted common names in *nabirds_names.npy*. If we want to classify *bird6* images, which use labels [0, 5], we must first map the labels to their matches in *NABirds*:

```
i = np.where(ytst==0)[0]; ytst[i] =   9   # Kestrel
i = np.where(ytst==1)[0]; ytst[i] =  15   # Pelican
i = np.where(ytst==2)[0]; ytst[i] =  31   # Belted Kingfisher
i = np.where(ytst==3)[0]; ytst[i] = 173   # Great Blue Heron
i = np.where(ytst==4)[0]; ytst[i] = 296   # Red-tailed Hawk
i = np.where(ytst==5)[0]; ytst[i] = 331   # Snowy Egret
```

The relabeled samples are now aligned with the output produced by each of the *NABirds* models.

The nearest-neighbor model relies solely on the CLIP embeddings, there is no training step. This is not the case for the MLP and SVM. Those models require training data. Additionally, to evaluate the model on *NABirds* itself, we must hold back some of those embeddings for test.

Most classes in *NABirds* contain 16 image embeddings, by design. The MLP and SVM classifiers use 10 of the 16 as training data and the remaining 6 as test data, which is the purpose of the function `TrainTestSplit`:

```
x = np.load("nabirds_features.npy")
y = np.load("nabirds_labels.npy")
xtrn,ytrn, xtst,ytst = TrainTestSplit(x,y)
```

Once split, the appropriate model object is instantiated

```
clf = MLPClassifier(hidden_layer_sizes=(nodes, nodes//2),
        max_iter=1000)
clf = SVC(C=1, kernel='linear')
```

then trained and applied to the test set,

```
clf.fit(xtrn, ytrn)
pred = clf.predict(xtst)
```

Now, let's dive deeper still into the results of this section in an effort to understand why the *bird6* dataset performed as it did.

7.3   Understanding the Models

Raw *NABirds* CLIP embeddings led to well-performing MLP and SVM models with 50 percent of classes achieving an accuracy of over 83 percent. However, the *bird6* embeddings were poor performers, maxing out in the low 50 percent range. Why?

The *NABirds* embeddings are based on chips extracted according to the supplied bounding box, a box that encompasses the bird in question. The *bird6* embeddings came from chips extracted by clicking on the image to define the largest square centered on that point. There is no assurance that the bird is entirely in the image. In other words, the *NABirds* embeddings are more complete representations of what CLIP makes of the bird. It stands to reason that more complete *bird6* images will do better.

The file *bird6_large_features.py* produces *bird6_large_features.npy* and associated label file from the full images used to build *bird6* in Chapter 4. The claim of the previous paragraph is tested by specifying `large` as the dataset when evaluating different models. For example,

```
> python3 nabirds_classifier.py large tmp
Test set accuracy: 0.5859, MCC: 0.5837
```

```
American Kestrel            (0.918)  (0.776)
American White Pelican      (0.966)  (0.931)
Belted Kingfisher           (0.957)  (0.609)
Great Blue Heron            (0.637)  (0.375)
Red-tailed Hawk             (0.105)  (0.023)
Snowy Egret                 (0.848)  (0.804)
```

Note that I've removed the confusion matrix and list of per-class assignments, retaining only the overall accuracy.

Performance on red-tails is still poor, but slightly better. Other classes are improved, often dramatically compared to the original *bird6* chips (the second column listed).

Improvement also appears with the top-level MLP and SVM classifiers:

```
Test set accuracy: 0.5732, MCC: 0.5536 (MLP)
Test set accuracy: 0.5909, MCC: 0.5609 (SVM)
```

Curiously, however, improvement in overall accuracy comes at the expense of specific classes as revealed by the per-class accuracies:

```
                        Cosine   MLP     SVM
American Kestrel        0.918    0.894   0.765
American White Pelican  0.966    0.828   0.931
Belted Kingfisher       0.957    0.609   0.783
Great Blue Heron        0.637    0.688   0.725
Red-tailed Hawk         0.105    0.226   0.346
Snowy Egret             0.848    0.609   0.435
```

Recall that the *bird6* test set is imbalanced and contains 133 red-tail samples. Therefore, even a modest improvement in classifying red-tails will lead to overall improvement even if other classes suffer.

In Chapter 6, CLIP embeddings produced a *bird6* classifier with better than 96 percent overall accuracy, yet, when using *NABirds* embeddings accuracy falls. The classifier of Chapter 6 used only the *bird6* labels, thereby forcing the classifier to select from among 6 text embeddings, not 404. It's reasonable to wonder if restricting the species list to what is likely to be present in the dataset is helpful.

All of the *bird6* images were taken in Adams County, Colorado. The eBird website (ebird.org) produces county-level data on bird sightings. I extracted the list of all species detected in the county over one year, 274 species, and used them to produce text embeddings for a cosine distance classifier. The code in *adams_classifier.py* mimics that of *nabirds_classifier.py* but is restricted to only those 274 species. Running the classifier against *bird6* and the *large* version validates our intuition. For example, the *large* images lead to:

```
Test set accuracy: 0.6288, MCC: 0.6301

American Kestrel            (0.918)
American White Pelican      (0.966)
Belted Kingfisher           (0.957)
Great Blue Heron            (0.812)
```

```
Red-tailed Hawk          (0.105)
Snowy Egret              (0.913)
```

Classes either maintain their previous accuracy or improve it: herons (64 to 81 percent) and egrets (85 to 91 percent).

The model of Chapter 6 forced assignment to one of the six classes to which we knew the test sample must belong. The Adams County model selects from among 274 possible birds, those we expect to be present in the region. Finally, the *NABirds* model chose from 404 species. Each successive level of restriction improved the results. In general, it is best to restrict models to the actual range of data they will encounter when used.

7.4   Generic Images and Text

We are now able to create generic bird classifiers for North American birds. Four classifiers are available:

1. *clip_text_generic.py*
2. *clip_image_generic.py*
3. *mlp_image_generic.py*
4. *svm_image_generic.py*

The first accepts a text prompt and matches it to common bird name embeddings. In other words, it matches text to text.

The final three classify arbitrary bird images. The first of these relies on the cosine distance from various image embeddings: images, means of images over each class or the common name. The last two use trained top-level MLPs or SVMs.

The code for each classifier is a variation on the themes of this chapter. Therefore, I leave it to you to read through the code. Each part will seem familiar.

7.4.1   Generic Text Classifier

Let's describe some birds and see what matches. A few command line arguments are expected:

```
clip_text_generic <top-n> <prompt> [<negate>]

   <top-n>  - capture the top-n best matches
   <prompt> - text describing the bird
   <negate> - 'negate' to keep the top-n _worst_ matches
```

The cosine distance ranks each common name embedding against the prompt embedding, therefore the first argument specifies the number of closest matches to return. The second argument is the prompt describing the bird (enclosed in double quotes). The final argument, which is optional, reverses the ranking to return the worst matches.

Let's take *clip_text_generic.py* out for a test drive:

```
> python3 clip_text_generic.py 6 "a black bird"
Prompt: a black bird
```

```
(0.170759)  Fish Crow
(0.209583)  Northwestern Crow
(0.210304)  American Crow
(0.245804)  Bald Eagle
(0.262989)  Brewer's Blackbird
(0.266897)  Great Cormorant
```

We asked for the six common bird name embeddings with the smallest cosine distance to the embedding of the prompt "a black bird." The crows, blackbird and cormorant are indeed black. Bald Eagles are not black birds, but much of an adult's body is brown, so call it a near miss. Notice the magnitude of the cosine distances: all are less than 0.27.

Let's repeat the query but negate the reply:

```
> python3 clip_text_generic.py 6 "a black bird" negate
Prompt: a black bird (negated)
(0.765179)  Pyrrhuloxia
(0.733851)  Yellow-bellied Sapsucker
(0.696377)  Eastern Wood-Pewee
(0.690813)  Chestnut-sided Warbler
(0.686399)  Green-tailed Towhee
(0.680931)  Western Wood-Pewee
```

None of the listed birds are black. Also, notice that the magnitude of the cosine distance is significantly greater than the closest matches.

If you run repeatedly using different colors, you'll find that the closest matches are often accurate, but also that some birds appear more often than we might expect them to. For example, Fish Crows and Bald Eagles appear more often than expected. Why? I'm not sure.

General groupings are also captured, at least partially:

```
> python3 clip_text_generic.py 6 "raptor"
Prompt: raptor
(0.241881)  Bald Eagle
(0.288747)  Golden Eagle
(0.291949)  Peregrine Falcon
(0.302490)  Wild Turkey
(0.303337)  Brant
(0.307521)  Red-tailed Hawk
```

Four of the six responses are indeed raptors, but why Wild Turkeys and Brants show up, and before Red-tailed Hawks, is a bit confusing. Recall, however, that we are using CLIP embeddings "out of the box" as it were. A true research project employing CLIP would likely fine-tune the model with bird lore to enhance its ability to encapsulate helpful distinguishing features.

Consider a final prompt seeking matches to "waterfowl" before switching to generic image classification:

```
> python3 clip_text_generic.py 6 "waterfowl"
Prompt: waterfowl
```

```
(0.213573)   Cackling Goose
(0.228752)   Mallard
(0.234081)   Snow Goose
(0.234558)   Wild Turkey
(0.239104)   Fish Crow
(0.242065)   Bald Eagle
```

Again, the first three matches are accurate, yet Fish Crow and Bald Eagle seem compelled to make an appearance.

7.4.2   Generic Image Classifiers

The first of the three generic image classifiers relies on the cosine distance between the embedding of the supplied image and *NABird* image or name embeddings. Arguments are:

```
clip_image_generic <top-n> <mode> <image> [<negate>]

   <top-n>   – capture the top-n best matches
   <mode>    – 'all', 'avg', 'navg', or 'name'
   <image>   – the bird picture to classify
   <negate>  – 'negate' to keep the top-n _worst_ matches
```

Most arguments are familiar. The mode, however, is new. If `all`, the embedding of the supplied image is compared to all 8877 *NABirds* embeddings, implying the top matches might repeat the same class name. The remaining modes use the per-class average embedding (avg), the common name embedding (name) or the per-class average and the common name (navg). This last mode is unlikely to offer much, if anything, over the per-class average, but I was curious, so I added it.

We need some examples to classify. You'll find sixteen in the *examples* directory; see Figure 7.2. Notice that the images are of arbitrary aspect ratio. The classifiers handle this by placing the image within a square image of the same size as the original's largest dimension, then resizing to 336×336 pixels for CLIP.

Let's classify the Killdeer using each of the modes, keeping the top 4 results:

```
Image examples/killdeer.png (all)
(0.070145)   Killdeer
(0.095909)   Killdeer
(0.099773)   Killdeer
(0.122133)   Semipalmated Plover

Image examples/killdeer.png (avg)
(0.094762)   Killdeer
(0.113758)   Semipalmated Plover
(0.141677)   Ruddy Turnstone
(0.156831)   Spotted Sandpiper
```

```
Image examples/killdeer.png (navg)
(0.095432)   Killdeer
(0.115377)   Semipalmated Plover
(0.144438)   Ruddy Turnstone
(0.158474)   Spotted Sandpiper

Image examples/killdeer.png (name)
(0.715897)   Killdeer
(0.742917)   Semipalmated Plover
(0.756584)   Least Sandpiper
(0.762317)   Ruddy Turnstone
```

All modes produce a correct top-1 classification. Comparing with all 8877 *NABirds* embeddings returned "Killdeer" as the top 3 matches. Notice that the smallest cosine distance was 0.07. The per-class averages, with or without the name embedding (navg), also hit with a small cosine distance below 0.1.

The name-only embedding, which compares the image embedding to the 404 *NABirds* common name embeddings, is correct, but the cosine distance is much greater, above 0.71.

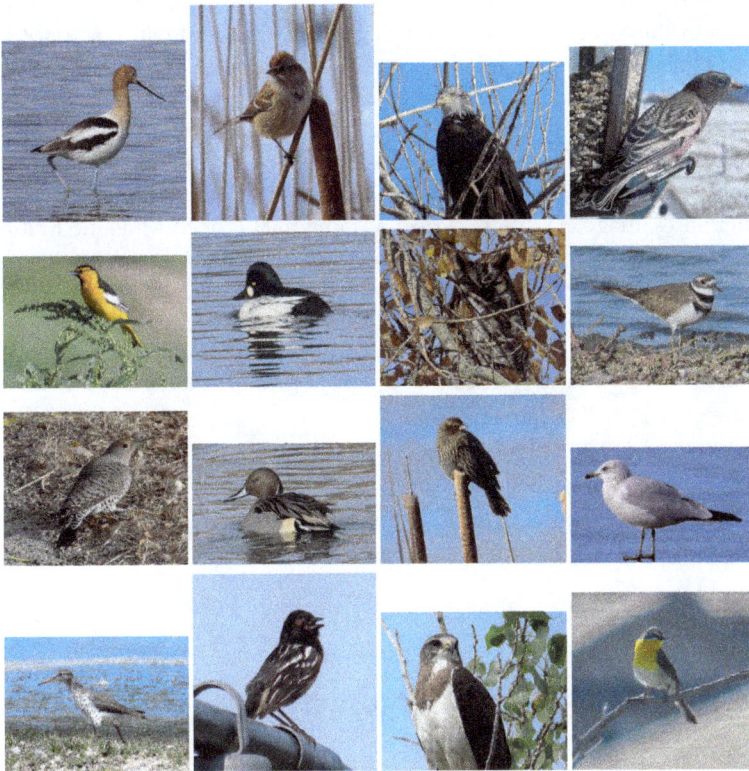

Figure 7.2 American Avocet, American Tree Sparrow, Bald Eagle, Brown-capped Rosy-Finch (row 1), Bullock's Oriole, Common Goldeneye, Great Horned Owl, Killdeer (row 2), Northern Flicker, Northern Pintail, Red-winged Blackbird, Ring-billed Gull (row 3), Spotted Sandpiper, Spotted Towhee, Swainson's Hawk, Yellow-breasted Chat (row 4)

The top-1 result for each of the 16 example images using avg mode returns:

```
American Avocet:          (0.077833)   American Avocet
American Tree Sparrow:    (0.125842)   American Tree Sparrow
Bald Eagle:               (0.122509)   Bald Eagle
Brown-capped Rosy-Finch:  (0.176523)   House Finch
Bullock's Oriole:         (0.089686)   Bullock's Oriole
Common Goldeneye:         (0.100851)   Barrow's Goldeneye
Great Horned Owl:         (0.075968)   Great Horned Owl
Killdeer:                 (0.094762)   Killdeer
Northern Flicker:         (0.090401)   Northern Flicker
Northern Pintail:         (0.081951)   Northern Pintail
Red-winged Blackbird:     (0.134311)   House Finch
Ring-billed Gull:         (0.038309)   Mew Gull
Spotted Sandpiper:        (0.085217)   Spotted Sandpiper
Spotted Towhee:           (0.154991)   Spotted Towhee
Swainson's Hawk:          (0.097006)   Swainson's Hawk
Yellow-breasted Chat:     (0.111546)   Yellow-breasted Chat
```

Four of the 16 are mislabeled using the top-1 result. However, of those four, the correct label appears in the top 2 or 3 with only the Brown-capped Rosy-Finch not appearing until 10th on the list. Confusing a Red-winged Blackbird with a House Finch seems a tad strange, but teasing out exactly why the CLIP embeddings led to such a result is likely extremely difficult, if not impossible.

ImageNet classifiers often report the top-5 accuracy. By that metric, the simple CLIP classifier delivers a respectable 15 out of 16 for an accuracy of about 94 percent.

The remaining image classifiers expect a trained MLP or SVM as the first argument, followed by the number of top results and the test image. Fortunately, *nabirds_mlp_classifier.py* and its SVM partner from earlier in the chapter stored their trained models for just this purpose.

For example, running the MLP and SVM generic classifiers on the Killdeer image gives:

```
> python3 mlp_image_generic.py mlp.pkl 4 examples/killdeer.png
Image examples/killdeer.png
   (0.99893)   Killdeer
   (0.00096)   Horned Lark
   (0.00004)   Solitary Sandpiper
   (0.00004)   Northern Bobwhite

> python3 svm_image_generic.py svm.pkl 4 examples/killdeer.png
Image examples/killdeer.png
   (1.00000)   Killdeer
   (0.99752)   Semipalmated Plover
   (0.99504)   Ruddy Turnstone
   (0.99257)   Spotted Sandpiper
```

It's important that we understand the numbers in each result. Scikit-Learn supplies a predict_proba method on the MLPClassifier class. This method returns the softmax output vector, which we understand is (with caution) interpretable as a set of probabilities. The numbers in the MLP output are the softmax probabilities sorted in decreasing order. Therefore, the MLP assigned a likelihood of 0.99893 to Killdeer with the next highest softmax likelihood of 0.00096 to Horned Lark. In other words, the MLP is extremely confident in its classification.

The SVM classifier was also correct, but care is required to understand the numerical score. The code uses the `decision_function` method of the SVC class (the trained SVM) to return a set of decision scores, one for each of the 404 possible output classes. This vector is scaled to fit into [0, 1] so that every response will begin with a score of 1.0 and go down from there. There is no easy way to interpret the magnitude difference between the scores other than decreasing scores imply less confidence in that label.

For example, applying the SVM to the difficult Brown-capped Rosy-Finch image, which appeared only in the top 10 when using cosine distance returns:

```
Image examples/brown-capped_rosy-finch.png
    (1.00000)   House Finch
    (1.00000)   Cassin's Finch
    (0.99752)   White-winged Crossbill
    (0.99503)   Pine Grosbeak
    (0.99255)   Red Crossbill
    (0.98510)   Black Rosy-Finch
    (0.98510)   Brown-headed Cowbird
    (0.98510)   Common Redpoll
    (0.98510)   Brown-capped Rosy-Finch
    (0.98013)   Lark Bunting
```

The correct label appears ninth on the list, but the difference between its score and the three above is less than 10^{-5}. The same image returns the correct label as fifth on the list when using the MLP classifier.

Applying the MLP and SVM classifiers to the 16 example images produces overall top-1 accuracies of 75 and 87.5 percent, respectively, the MLP matching the top-1 accuracy of the cosine distance classifier. Switch to top-5 accuracy, and the MLP is perfect, while the SVM hits 15 of the 16 to match the cosine distance top-5 accuracy.

We learned in Chapter 5 that it's possible to ensemble individual classifiers to gain still more accuracy. Can we do the same for the three image classifiers? Definitely.

The nearest-neighbor model ranks class labels by the cosine distance. The MLP uses a softmax vector of per-class likelihoods, and the SVM ranks via decision function value. Many options exist to combine these disparate rankings, the simplest being voting among the top-1 labels.

The file *ensemble_generic.py* does just that. The selected label is printed along with the label chosen by each of the three classifiers. Let's apply the code to the images in the *example* directory.

Of the 16 examples, 11 are correctly identified with each classifier selecting the same label, 3 are mislabeled, and 2 are correctly labeled by majority vote:

```
** Brown-capped Rosy-Finch:
        House Finch         (cosine)
        House Finch         (MLP)
        House Finch         (SVM)
```

```
** Common Goldeneye
        Barrow's Goldeneye
        Barrow's Goldeneye
        Barrow's Goldeneye
** Ring-billed Gull
        Mew Gull
        Glaucous-winged Gull
        Ring-billed Gull
    Red-winged Blackbird:
        House Finch
        Red-winged Blackbird
        Red-winged Blackbird
    Spotted Towhee
        Spotted Towhee
        Lark Bunting
        Spotted Towhee
```

All of the models were equally wrong about the Brown-capped Rosy-Finch and the Common Goldeneye. Notice, however, that one of the models (the SVM) correctly identified the Ring-billed Gull, but since each model selected a different bird, the actual output label is chosen at random with Glaucous-winged Gull on my first run and Ring-billed Gull on the second.

This quick ensembling experiment shows promise and might become an important part of a research project seeking a generic bird image classifier based on CLIP (or other) embeddings. The critical question is how to combine the measures produced by the separate models, beyond simple voting.

7.5   Discussion

The chapter's goal was to develop generic bird classifiers using CLIP features. We were successful, perhaps more so that we had a right to be given the simplicity of our approach. We benefited greatly from the wealth of information buried in CLIP embeddings and the massive amount of computing (and human) resources that went into developing the CLIP model. Such success wouldn't have been deemed likely much before 2022.

We also learned that when possible, we should restrict the set of possible classes to what we need for the task at hand. If we know the model will see only one of a dozen or so bird species, then comparing against those embeddings (common name, average over some known set of images per class, etc.) is all that should be done to maximize accuracy. Still, the full *NABirds* dataset, all 404 classes, produced results where 50 percent of the classes achieved over 83 percent accuracy, a result that would have been celebrated not long ago.

Restricting the set of known classes has a potential downside, however. What happens when a rare bird shows up in an area where it isn't expected? Or, due to climate change, begins making more frequent appearances in regions outside its historic range? As we learned in earlier chapters, a basic model will assign the novel input to a known class unless care is taken to implement a decision threshold applicable to a cosine distance classifier. However, selecting a single threshold based on cosine distance will be tricky.

It's well understood that random vectors in high-dimensional spaces are, on average, nearly orthogonal, implying many cosine distance calculations for many valid hits will be in the vicinity of 1. That said, anecdotal observation while building the experiments of this chapter hints that a significant fraction of the best matches will produce cosine distances of less than 0.8 or even 0.76. Consider the observation a rule of thumb at best.

In the end, the experiments of this chapter point the way for those seeking more comprehensive models, models that might be deployed to autonomously monitor birds in a specific region or as an online tool for the general public to use.

Classification identifies the bird in the image but does not tell us *where* it is in the image. For that, we need localization and detection, the subject of the next chapter.

8. Detection

The models we've explored to this point in the book fall under the umbrella of *object classification*. The models tell us whether the picture contains a Golden Eagle or a Snowy Egret, but they do not help us locate the bird in the image. *Object localization*, on the other hand, marks the position of objects in the image. Put the two together, and you have *object detection*: the *where* (localization) and *what* (classification) of the object. Detection is the next logical step beyond classification, and essential for many deep learning tasks. Fortunately, it's an area with a long history (relatively speaking for such a young science), so we have options.

The chapter begins with a review of what I'm calling the *detection hierarchy*, a series of increasingly detailed evaluations and understandings of the content of an image. The hierarchy developed over time, with each increase in knowledge following logically from the previous.

Implementation of each step in the hierarchy exceeds our mandate. Therefore, we'll experiment here with only two approaches. The first is a poor person's implementation of the original sliding window technique. Going old-school lets us apply powerful, information-rich CLIP embeddings to a simple detection task.

The second set of experiments focuses instead on *fully convolutional networks (FCN)*, themselves a modification of the CNNs used for object classification. The modification removes restrictive dense layers and replaces them with convolutional layers. Doing so eliminates any baked-in dependence on input image size, allowing us to apply the network to arbitrary images from which we can map where and how strongly the model believes a target bird is. I think of this chapter as a quick dip in the pool. Serious detection projects should contemplate the other detection techniques we merely discuss here.

8.1   The Detection Hierarchy

The detection hierarchy flows from most basic to most advanced in terms of what the method can tell us about the contents of an image. Stepping from least-informative to most-informative approach reveals the hierarchy:

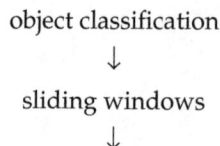

object classification

↓

sliding windows

↓

fully convolutional networks

↓

bounding box regression

↓

semantic segmentation

↓

instance segmentation

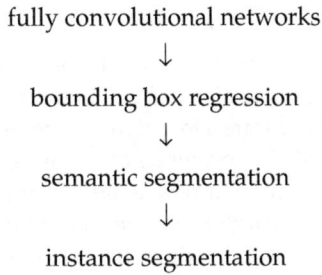

Object classification is the least informative approach: it tells us that something is somewhere within an image, but offers (seemingly) no information on where, or indeed that there might be other things of interest (i.e., birds) in the image.

I added "seemingly" because of the way we implemented the generic image classifiers of Chapter 7 does reveal additional information about the image, but not in a way that is particularly helpful without already knowing the image content.

The *examples* directory contains the images we'll use in this chapter. One of them is *kestrel_flicker.png*, an image containing exactly two birds and little else. The kestrel we are familiar with; the other is a Northern Flicker, which is particularly common where I live. We'll view this image later in the chapter, but for now, consider what the generic image classifiers of Chapter 7 make of it:

```
> python3 clip_image_generic.py 5 avg kestrel_flicker.png
Image /kestrel_flicker.png (avg)
    (0.155267)   Northern Flicker
    (0.178279)   Gila Woodpecker
    (0.202901)   Golden-fronted Woodpecker
    (0.203680)   Ladder-backed Woodpecker
    (0.213733)   American Kestrel

> python3 mlp_image_generic.py mlp.pkl 5 kestrel_flicker.png
Image kestrel_flicker.png
    (0.99995)   Northern Flicker
    (0.00002)   Cactus Wren
    (0.00002)   Gila Woodpecker
    (0.00000)   Inca Dove
    (0.00000)   American Kestrel
```

The CLIP embedding classifier, using per-class average embedding vectors, captures the flicker but also returns the kestrel in the top five closest cosine distances. The MLP-based classifier, which we will use in this chapter without retraining, does the same, showing softmax output. The model is exceedingly confident about its flicker assignment, but even through the softmax probability is virtually zero still returns the kestrel when demanding the top five results.

The image classifiers do in some way detect the two birds, but without knowing ahead of time that there are indeed two, and only two, birds in the image, and that one of them is a kestrel, we would not be able to use these results reliably. We must do something to localize the specific instances. Sliding windows are a natural next thought.

8.1.1 Classification with Sliding Windows

The image classifiers of Chapter 7 consider the entire input as a whole before returning the best result based on minimum cosine distance to a known CLIP embedding or a softmax output probability. If we want to locate a target in a larger input image, it makes sense to apply the image classifier repeatedly to subsets of the image. In that way, we build up an output set of likelihoods each associated with a particular portion of the input image. In other words, we convolve the classifier itself over the input image by sliding a window of the size the image classifier expects, position by position.

Sliding windows are the first technique applied to convolutional models for images exceeding the dimensions the CNN expects: grab a window, classify it, keep the output label and softmax value, step over some number of pixels or fraction of the window size, and repeat until the entire image has been processed.

Sliding windows succeed, and we'll test them later in the chapter as a way for us to keep working with CLIP embeddings, but it doesn't take too long to realize that they are exceedingly slow and computationally expensive. Fine-grained localization of a target bird is possible at the expense of many, many CPU or GPU cycles. Increasing the step size moves more quickly through the image, but degrades localization performance. It's entirely possible to do a "quick" scan of the image, then focus on the most promising areas, but in the end, excessive computation is still required.

For example, assume a CNN trained on 64×64 pixel inputs. Further, assume we want to pass this CNN over an input image that is 640×1280 pixels (height and width) using a step size of 32 pixels, half the input window width. This means that the first call to the CNN will use columns 0 through 63 of the larger image with the second using columns 32 through 95, and so on. Therefore, 39 calls are required to process the first "row" of the larger image. Stepping down by 32 pixels to process the second implies another 39. There are 19 such steps down the image to process all of it. In the end, then, to localize targets within the 640×1280 image requires $19 \times 39 = 741$ calls to the CNN. Such is possible, and perhaps not too terrible if the CNN is efficiently implemented on a GPU, but localization isn't precise. Step by one pixel, and the number of calls skyrockets to a prohibitive 702,209 for a single image! Computer scientists refer to this effect as "combinatorial explosion."

Sliding windows are a brute-force solution. A clever observation about what restricts CNNs to fixed-sized inputs, and how to bypass that restriction, enabled a step down to the next level in the hierarchy: fully convolutional CNNs without dense layers.

8.1.2 Fully Convolutional Networks

The CNNs we've explored so far, LeNet-5, VGG8 and ResNet-18, expect a fixed-sized input image of either 32×32 or 64×64 pixels. Why? What is it in the architecture of the model that forces such a restriction? The answer to that question leads to the next historic detection improvement. Let's work with LeNet-5. The Keras code for our implementation is:

```
inp = Input(input_shape)
_   = Conv2D(6, (3,3))(inp)
_   = BatchNormalization()(_)
_   = ReLU()(_)
_   = MaxPooling2D((2,2))(_)
_   = Conv2D(16, (3,3))(_)
```

```
_  = BatchNormalization()(_)
_  = ReLU()(_)
_  = MaxPooling2D((2,2))(_)
_  = Conv2D(120, (3,3))(_)
_  = BatchNormalization()(_)
_  = ReLU()(_)
_  = Flatten()(_)
_  = Dense(84)(_)
_  = BatchNormalization()(_)
_  = ReLU()(_)
_  = Dropout(0.5)(_)
_  = Dense(num_classes)(_)
outp = Softmax()(_)
```

We are already familiar with the structure, but let's review. The model is built, layer by layer, beginning with the Input, which expects a tuple, (H, W, C), for height (H), width (W) and number of channels (C), 1 for grayscale and 3 for RGB.

Which of the layers in the model, when trained, have a dependence on the input size, meaning the tuple (H, W, C)? The convolutional layers learn filters with particular kernel sizes, here 3×3. Nothing in the convolutional layer depends on the height or width of the input to the layer, it will happily convolve over any input size to produce an output with the same height and width (padding='same') or slightly reduced if the default padding of 'valid' is selected. The LeNet-5 implementation above uses the default, implying an input of (H, W) becomes $(H - 2, W - 2)$ on output because the filter kernels are 3×3.

The ReLU, BatchNormalization, and MaxPooling2D layers are similarly indifferent to height and width. Therefore, most of the model can operate over arbitrary input sizes.

The Dense layers, however, are a problem. They learn weight matrices mapping their input vectors to their output vectors, and those depend critically on the size of that vector. The dense layers prevent the model from operating over arbitrarily large input images. Train for 64×64, and you must supply 64×64 as input.

Replacing the dense layers with carefully selected convolutional layers would remove the input size restriction. But, how? The short answer is to replace dense layers with convolutional layers using a kernel size matching the height and width of the input to the layer and with as many filters as there are nodes in the dense layer itself. If the input to the layer is a vector, as in the second Dense layer of the LeNet-5 model, use a 1×1 kernel. If the input is of a different height and width, make the kernel that height and width. Convolving an $n \times n$ kernel over an $n \times n$ input produces a single output value. Applying k such filters with $n \times n$ kernels produces k output values, a $1 \times 1 \times k$ tensor matching the k nodes of the dense layer being replaced.

For example, the second LeNet-5 dense layer:

```
_  = Dense(num_classes)(_)
```

becomes

```
_  = Conv2D(num_classes, (1,1))(_)
```

to produce an output tensor of `num_classes` channels with a height and width related to the height and width of the input image and the spatial effect induced by the network layers. Later in the chapter, we'll explore the association between this output size and the input image to map the model's now spatially distributed output to input image pixel space.

The fully convolutional network trick was first presented by UC Berkeley researchers Jonathan Long, Evan Shelhamer and Trevor Darrell in 2015. With it, the unrealistic compute necessary for fine-grained localization via sliding windows was eliminated. The chapter's experiments demonstrate decisively that fully convolutional models are significantly faster than any sliding window implementation could ever be.

We'll convert a standard CNN into a fully convolutional model in the experiments. For now, let's proceed to the next innovation in detection.

8.1.3   Bounding Box Regression

A standard localization technique wraps detections in a *bounding box*, a rectangle just large enough to enclose the object of interest. The next innovation, at least conceptually though not temporally, beyond fully convolutional models involves learning both a label and the coordinates of a bounding box for each object in the image recognized by the network. Many approaches exist, but a typical network output, assuming a single detection, might be a vector specifying a softmax over the possible class labels along with the lower-left corner of the bounding box followed by its height and width. Bounding box coordinates are often normalized to [0, 1], implying the output given as a fraction of the input image's height and width.

For example, assume we have a trained network capable of classifying a single bird from among five possible types along with the bounding box localizing the bird. In that case, a possible output vector might be:

```
0.2 0.0 0.1 0.1 0.6 0.3 0.4 0.1 0.2
```

The first five numbers are a softmax vector identifying class 4 as the label. The next four numbers are the lower-left corner of the bounding box followed by the height and width, all as fractions of the input image height and width. Note that such a model typically involves two pathways: one leading to a cross-entropy loss, the softmax output, and another leading to a sigmoid loss, the bounding box output. The convolutional and pooling portion is learned in common between the two outputs (often referred to as *heads*).

Practical application of bounding box models often employs the YOLO ("you only look once") model. YOLO is fast and labels multiple instances of known object classes. A pretrained version is downloadable from `https://pjreddie.com/darknet/yolo/` and runs on a standard CPU-only machine to classify objects in arbitrary images.

Figure 8.1 presents YOLO output for two of the "duck" images in the *examples* directory. The bounding boxes are all labeled "bird" since YOLO does not know how to distinguish between species. Notice that most of the birds are captured, but not all. Fine-tuning YOLO or a similar model like RetinaNet would be required to expand the model to specific bird classes.

8.1.4   Semantic Segmentation

Image segmentation labels object pixels to become next-level detection beyond bounding boxes. The goal is a mask revealing all the pixels associated with the object.

Figure 8.1 YOLO output for two of the "duck" example images.

This approach is known as *semantic segmentation*, where every pixel associated with a particular class receives the same label. Architecture options include modified vision transformers, a current state-of-the-art approach, and U-Net, a common choice in medical imaging. For example, semantic segmentation of the lower image in Figure 8.1, assuming only the "bird" label, would mark each pixel of the three birds as belonging to that class. This implies that the model's output is a mask image of the same size as the input where each pixel receives a class label. In practice, if the model recognizes N classes, the output becomes a $H \times W \times N$ binary tensor for input image dimensions $H \times W$ and each channel a binary mask indicating image pixels assigned to that class (1) or not (0).

Consider Figure 8.2. It shows a possible semantic segmentation output for the "bird" class. The mask marks pixels identified as "bird" but without grouping into individual birds as far as the model is concerned. In other words, the classification is pixel-level only, that sets of pixels correspond to specific birds is not included in semantic segmentation.

If that's the case, why use the word "semantic?" Region growing, a traditional image segmentation technique, groups pixels according to a set of conditions, such as the appearance of a sudden gradient in the image (a change of color or a transition from light to dark). The algorithm has no understanding of what the object might be or why particular pixels are to be associated with that segmentation.

Semantic segmentation, on the other hand, relies on the model's learned understanding (however appropriately defined) of the content of the image and from that comes the per-pixel labels.

Figure 8.2 Semantic segmentation labels pixels of the same class.

Segmentation followed by a traditional technique like region labeling can serve to identify specific connected regions in the mask. For example, if applied to Figure 8.2, such an algorithm would easily distinguish the three birds. However, if the birds are adjacent or overlapping, it is not possible (or at least easy) to separate them based on semantic segmentation masks alone. For that, instance segmentation is required.

8.1.5   Instance Segmentation

Semantic segmentation labels individual pixels. Bounding boxes identify instances of the object. Instance segmentation combines the two approaches to label pixels belonging to individual instances of known classes.

Figure 8.3 illustrates an instance segmentation of Figure 8.2. The same pixels are located along with knowledge about which pixels are associated with which birds. Several different models implement instance segmentation with Mask R-CNN as a standard example.

Mask R-CNN by He, Gkioxari, Dollár, Ross Girshick (2017) is a multi-stage model building on Faster R-CNN by adding an additional stage to predict object masks. A backbone model predicts image features; think of the embedding vectors of Chapter 6. The next stage then uses the features to predict bounding boxes. Finally, a separate model head simultaneously predicts object masks for each bounding box. Joint training of the bounding box prediction head and mask head leads to a model capable of both, thereby producing instance-level object segmentations.

Instance segmentation is the end goal of our detection hierarchy beginning with simple "There's a bird in this image" classification all the way to "There are three birds in this image, here they are along with their segmentation masks."

Experimenting with bounding box, semantic, and instance segmentation is beyond what we can implement here, but this section should be read as a primer leading to further research for readers interested in more advanced detection approaches.

We learned in Chapter 7 that CLIP embeddings are a powerful tool for building classifiers with minimal or no data. We'd like to use these embeddings for detection as well. Advanced projects will no doubt combine the CLIP model with architectures akin to YOLO or Mask R-CNN, but the experiments of the following section harken back to early days merging CLIP embeddings with sliding windows as a quick-and-dirty approach to (crude) localization.

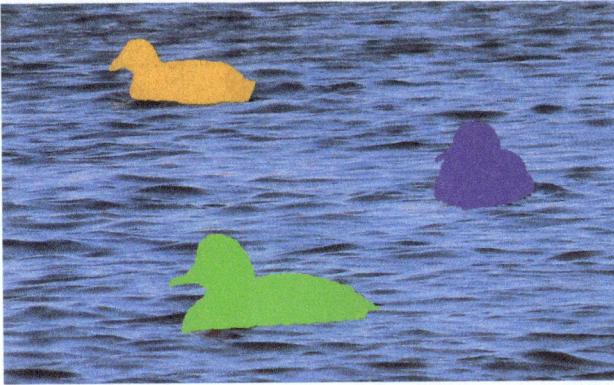

Figure 8.3 Instance segmentation groups pixels of the same class into separate instances.

8.2   Experiment: CLIP Embeddings

Let's expand the CLIP-based generic bird classifier of Chapter 7 to work with sliding windows. The expansion is straightforward: replace the single image with a set of smaller images covering the entire image, then classify each image as before using the cosine distance between bird name embeddings and the CLIP embedding of the smaller image. The code is in *clip_detection.py*. As always, I do recommend reading the code before continuing.

CLIP prefers 336×336 pixel inputs. We'll generate candidate windows by first resizing the input image to match a user-supplied aspect ratio corresponding to a given number of rows and columns. The initial set of subimages will be non-overlapping followed by a second non-overlapping set offset by half a window length. This approach may read as complex, but the implementation is conceptually straightforward. The user-specified rows and columns should be reasonably close to the image's actual aspect ratio to avoid excessive distortion of the input image.

For example, if the input image is $H \times W = 900 \times 1000$ pixels, then select three rows and three columns since $3/3 = 1$ and $900/1000 = 0.9$. However, if the image is 451×1000, then choose two rows and four columns for an aspect ratio of 0.5 which is close to the true aspect ratio of 0.451.

Figure 8.4 shows an example image rescaled to match a 3×3 aspect ratio on the left. The image on the right illustrates the set of sliding windows evaluated by the code. There are nine non-overlapping followed by an additional four offset from the initial set for a total of 13 evaluations. Notice that the windows are numbered. The code's output uses this numbering to identify bird detections. The top-1 bird for each subimage, assuming the cosine distance is below a user-supplied threshold, is output as a detection.

Let's give the code a try with the image in Figure 8.4, which you'll find in the *examples* directory as *snowy_avocet.png*:

```
> python3 clip_detection.py 0.71 3 3 examples/snowy_avocet.png
    tmp
```

The code expects five arguments: a cosine distance threshold, the number of rows and columns, the image to classify and an output directory name.

Figure 8.4 A scaled input image and the subimages selected for CLIP classification.

After about a minute, the code generates some text:

```
Chip classification (3 rows, 3 columns):
   2: (0.70294)   Snowy Egret
   4: (0.69613)   American Avocet
   7: (0.70513)   American Avocet
  10: (0.70225)   Snowy Egret
  11: (0.70530)   American Avocet
```

and several files in the *tmp* directory:

```
command_line.txt
console.txt
original_image.png
overlay_image.png
scaled_image.png
```

The text indicates the number of rows and columns, then lists subimages with a top-1 classification result below the supplied threshold of 0.71. The experiments of Chapter 7 taught us that many cosine distances between images and bird names land in the region between 0.7 and 0.8; therefore, a threshold of 0.71 tells the code to keep only those results with a reasonable likelihood of being a true match.

The numbers correspond to the windows on the right in Figure 8.4. Windows are numbered left to right, top to bottom regardless of the number of rows or columns. In this example, the first subimage with a result below the threshold is number 2, the upper right portion of the input image with a cosine distance of 0.703 and the label "Snowy Egret." The remaining detections follow in kind. All detections are either Snowy Egret or American Avocet, as we expect. Therefore, we have some confidence that the code is working.

The output directory contains two text files noting the command line and all text output to the console. There are also three image files: the original image, the image as rescaled for window extraction, and an overlay image. The overlay image contains a grayscale version of the scaled image with output regions highlighted in red using an alpha value based on the cosine distance; see Figure 8.5.

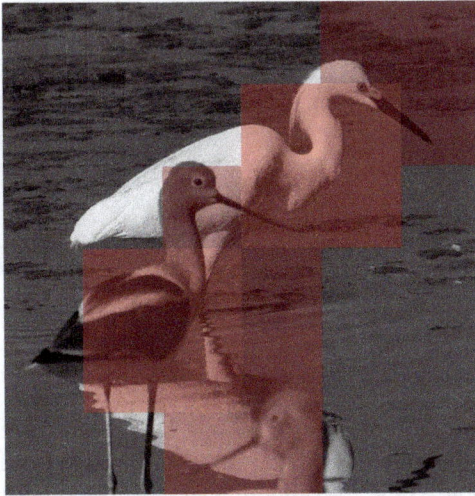

Figure 8.5 The overlay image marking CLIP detections.

Comparing the window numbers in Figure 8.4 with the text generated by the code and the overlay image in Figure 8.5 indicates that the output is as expected: window 2 (upper left) contains the Snowy Egret's head, while window 10 contains its neck. Both are marked as "Snowy Egret." Similarly, windows 4 and 7 contain the avocets' heads with window 11 the body.

8.2.1 Additional Examples

The CLIP detector is, to be honest, exceedingly minimal. However, it does accomplish its goal. Here are few more images from the *examples* directory:

```
> python3 clip_detection.py 0.72 2 4 gull_crow.png tmp
Chip classification (2 rows, 4 columns):
  2: (0.71339)  Common Raven
  5: (0.71327)  Ring-billed Gull
  8: (0.70062)  Ring-billed Gull
  9: (0.71691)  Ring-billed Gull

> python3 clip_detection.py 0.72 3 3 kestrel_flicker.png tmp
Chip classification (3 rows, 3 columns):
  2: (0.70712)  American Kestrel
  6: (0.67782)  Northern Flicker
 10: (0.71322)  American Kestrel
 11: (0.69202)  Northern Flicker

> python3 clip_detection.py 0.7 2 3 junco.png tmp
Chip classification (2 rows, 3 columns):
  1: (0.69717)  Dark-eyed Junco
  6: (0.68821)  Dark-eyed Junco
  7: (0.69616)  Dark-eyed Junco
```

with corresponding overlay images in Figure 8.6

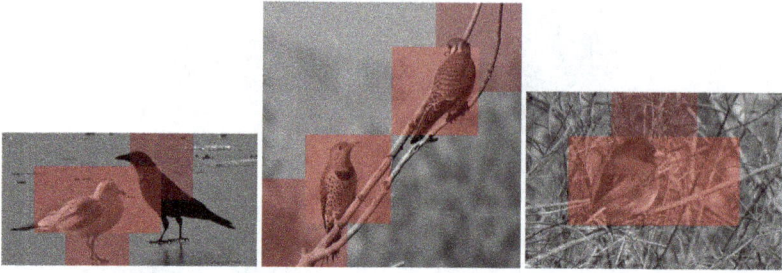

Figure 8.6 More CLIP detection examples

The examples are, admittedly, cherry-picked, but not so exceptional as to disprove the correct functioning of the code. The first example mistakenly labeled the crow as a raven, which is understandable as the two are quite similar. The other two examples correctly identified the target birds with cosine distances well below the supplied threshold. My experimentation hints that a distance below about 0.7 is typically correct, or at least close. Set the threshold to 2, the maximum cosine distance, to force selection for every window, whether a bird is present or not. For example, the kestrel and flicker image with a threshold of 2 returns:

```
Chip classification (3 rows, 3 columns):
    0: (0.74671)  Cliff Swallow
    1: (0.72489)  Vesper Sparrow
    2: (0.70712)  American Kestrel
    3: (0.72681)  Northern Flicker
    4: (0.72138)  Bobolink
    5: (0.72822)  American Kestrel
    6: (0.67782)  Northern Flicker
    7: (0.73050)  Northern Flicker
    8: (0.74824)  Swamp Sparrow
    9: (0.74974)  White-throated Swift
   10: (0.71322)  American Kestrel
   11: (0.69202)  Northern Flicker
   12: (0.73986)  Dickcissel
```

A high threshold reveals other windows where the top-1 label matches the bird along with the four present in Figure 8.6 (2, 6, 10 and 11). The high threshold ensures that windows 0, 8 and 9, which contain nothing but background, also select a bird, but the corresponding cosine distances are above the ad hoc 0.7 or so threshold, leading to convincing labels.

If the maximum cosine distance is 2.0 and orthogonal vectors have a distance of 1.0, why do most of the measured distances hover near 0.7? I suspect two factors are at play. First, in high-dimensional spaces, random vectors are generally almost orthogonal, which intuitively makes sense and implies cosine distances of about 1.0. The code in *random_vec. py* demonstrates this with histograms for random vectors in [− 0.5, 0.5) for a given number of dimensions. I recommend giving it a go.

Second, even windows without birds are not truly empty, they contain background information extracted from the photo, and it is reasonable to expect that CLIP learned to represent that background information in a way where there is still some essence (for lack of a better word) that is captured in the mapping of the bird name leading to top-1 matches, by chance, that are not significantly above the expected 0.7 threshold.

Detection quality is a function of the number of windows and what part of a bird appears in the window. Experimentation is needed to select a number of windows that match the image aspect ratio and the size of the birds in the image.

Let's quickly review *clip_detection.py* before moving on.

8.2.2   Reviewing the Code

The generic classifiers of Chapter 7 generated CLIP embeddings of the input image, then searched for the closest match to the name embedding vector. The same approach applies to sliding windows, with each chip a new image to classify. In code, this becomes gathering the chips (subimage) for each window, then sequentially classifying them:

```
simage, chips = Chips(image, nr, nc)
res = []
for chip in chips:
    img = Image.fromarray(chip)
    im = preprocess(img).unsqueeze(0).to(device)
    with torch.no_grad():
        features = model.encode_image(im)
    image_features = features.cpu().numpy().squeeze()

    scores = []
    for i in range(len(x)):
        scores.append(Cosine(x[i], image_features))
    scores = np.array(scores)
    order = np.argsort(scores)
    score = scores[order][0]
    label = y[order][0]
    res.append((score, label))
```

The code compares the common bird name embedding (x) with the image embedding (image_features) and selects the smallest cosine distance as the top-1 label. Thresholding is applied after all chips have been classified.

The Chips function resizes the input image so that each chip is exactly 336 × 336 pixels, then partitions the resized image into a collection, each of which is classified to output a detection if the cosine distance is below the user-supplied threshold. The Chips function is a brief exercise in image manipulation, but worth discussing.

The function consists of two sets of nested loops. The first runs over the given number of rows and columns, thereby dividing the image into disjoint chips. The second is the same but runs over one fewer row and column and applies a half-chip width offset. The resized image and collection of chips are then returned. The remainder of the code in *clip_detection.py* handles output.

8.2.3 Using an MLP

The code in *clip_detection.py* relies on the CLIP embeddings of common bird names. It does not use the embeddings derived from the *NABirds* dataset. In Chapter 7, we created a generic classifier from those embeddings by training a simple MLP. There is no reason why we cannot use that MLP with sliding windows. The code in *mlp_detection.py* does just that.

The code expects a trained MLP. Any such model trained in Chapter 7 will suffice (see *mlp.pkl*). It also expects a threshold, but here, the threshold has a different meaning. The MLP generates a softmax vector representing the model's belief over the full 404 bird classes it knows. The threshold is applied by looking for softmax values *above* the threshold, not below as with the cosine distance.

Let's use the MLP to detect the birds in the kestrel and flicker image in Figure 8.6:

```
> python3 mlp_detection.py mlp.pkl 0.96 3 3 kestrel_flicker.png
     tmp
Chip classification (3 rows, 3 columns):
  2: (0.99099)  American Kestrel
  3: (0.99988)  Northern Flicker
  6: (0.99934)  Northern Flicker
 11: (1.00000)  Northern Flicker
 12: (0.97403)  Cliff Swallow
```

Both birds were detected, but a false positive is present in window 12. Neural networks are notoriously overconfident, which is seen in the extremely high likelihoods given for the birds actually present in the image. The takeaway message is to use high thresholds with the MLP. Notice that the false positive produced a softmax value of 0.97 compared to the 0.99 and greater produced by true positives.

Feel free to experiment with the MLP detector. The softmax threshold has a more intuitive meaning than an ad hoc cosine distance threshold, but the MLP itself isn't calibrated and therefore overconfident. It's possible to calibrate the MLP to turn the softmax output into a true probability of class assignment (search for *Platt scaling*), but I don't recommend the exercise because of the inherent localization limitation of sliding windows.

Let's try to do better, at least in terms of localization and execution speed. For that, we want a fully convolutional network. Losing CLIP embeddings is the price we pay, but for specific purposes, say a detector for a target species, the price might be worth it.

8.3 Experiment: Fully Convolutional Networks

Let's construct a fully convolutional network to detect American Kestrels. Why kestrels? Because the bird is common in my area as I write this, I have many images to build a dataset. And therein lies our first challenge: creating the dataset.

The *bird6* dataset from earlier chapters contains kestrels, but we cannot use those images as they are. Fully convolutional networks convolve over the input image, implying that many of the inputs scanned will contain not all of the target class, but merely a portion of the target. Therefore, we want the model to respond to inputs that contain portions of the target. Not including such inputs in the training set conditions the model to react to only a very narrow range of cases where the input contains virtually all of the target.

Our goal is a binary model that assigns each pixel of the input image to one of two classes: background (0) or kestrel (1). I've already constructed the dataset; you'll find it in the *data* directory with our standard naming convention:

```
kestrel_xtrain.npy
kestrel_ytrain.npy
kestrel_xtest.npy
kestrel_ytest.npy
```

We have the dataset, but it's worth reviewing the process that created it. Recall that proper dataset construction is one of deep learning's essential components (and least sexy task). However, before we build the dataset, we must determine the structure of the fully convolutional model.

Fully convolutional models are adaptations of existing CNNs. Therefore, we need a base model architecture to train and from that construct the fully convolutional version. I selected a base architecture derived from LeNet-5. It expects a 64×64 RGB inputs to produce a 2-class softmax output: kestrel or not. We'll review the architecture later in this section along with the fully convolutional version. We train the base model on 64×64 inputs, but we must use inputs that include partial kestrel imagery, not merely the entire bird or a significant portion of the bird.

Imagine a 64×64 sliding window moving over a larger image from left to right. There is a kestrel in the path that the window will approach from the left. At first, only a tiny part of the kestrel will be on the right side of the window, but as the window moves further to the right, more and more of the kestrel will be captured. Later, as the window moves to the right, the kestrel will fade to the left until no longer in the window. We want every (most) of those windows containing the kestrel to be captured and marked as a target, not just the window over the majority of the bird.

So, where are we, then? We have a base model that we will train with 64×64 input images, but we also know that we'll hack the model to be fully convolutional after training and that the fully convolutional approach requires us to include training images that contain portions of the target, not just the target.

The *bird6* dataset relied on *chipper.py*, a graphical program to select image chips from larger images. I modified the program to extract multiple chips from the selected location by sliding over a region near where the user clicks. You'll find the code in *chipper2.py*.

For example, Figure 8.7 shows a kestrel image on the left with a red cross marking where I clicked. The program then automatically generated the 49 chips on the left, each 64×64 pixels. I ran *chipper2.py* over the *bird6* training images to produce a collection of 2254 image chips. The number and spacing of the chips are suitable for the experiment, but if adapting this technique to new data, I suggest perhaps expanding the size of the region exported and increasing the step size between chips from 1 pixel to 2 or even 4.

Our goal is a binary classifier, so we need negative examples, chips that are not kestrels. For that, I used *background_chips.py* to extract 100 chips at random from a collection of images without kestrels. Some careful consideration is required.

I selected images that *do* contain kestrels, but only subsets where the kestrel was not present. The rationale is that such background images are typical of places where kestrels are seen and therefore likely to be present in the kinds of images encountered by the model in the wild. This approach led to 2300 background chips, bringing the total number of images in the base kestrel dataset to $2254 + 2300 = 4554$ with 3643 in the training set and the remainder held back as a base model test set; see *build_kestrel.py*.

Figure 8.7 A kestrel image (left) and extracted chips (right).

Three tasks remain: train and test the base model, map the base model onto a fully convolutional (FCN) version and apply the FCN model to real-world images.

8.3.1 Training the Base Model

The code to train the base model is in *kestrel_base.py*. It follows the format used consistently to this point in the book: load the train and test data, construct the model, train and then evaluate. The command line I used was:

```
> kestrel_base.py 48 14 results/kestrel_base
```

Training takes only a few minutes to produce

```
[[453    2]
 [  0 456]]
Test set accuracy: 0.9978, MCC: 0.9956
```

informing us that the base model has performed quite nicely on the test set with a nearly perfect score. We should be suspicious of such a high score, to be honest, but the model was asked to differentiate between image chips containing kestrel content and fairly bland background images of sky and twigs, implying that the results might be legitimate – we'll know better when we use the model in the wild.

It's important to remember that the trained base model is no different than the models we used in earlier chapters. It's a basic CNN with dense layers restricting it to 64×64 inputs. The next task is constructing a fully convolutional version and populating it with base model weights.

8.3.2 Creating the Fully Convolutional Model

Transforming the trained base model into a fully convolutional model requires defining the fully convolutional architecture. First, by replacing base model dense layers with appropriately defined convolutional layers, then, second, by copying base model weights to their fully convolutional model versions modifying the shape of the weights as necessary for the new convolutional layers.

Creating the FCN Model

The code in *kestrel_fcn.py* constructs the fully convolutional model from the trained base model:

```
> python3 kestrel_fcn.py results/kestrel_base/model.keras
                      kestrel_fcn.keras
```

The first argument is the base model, and the second is the new fully convolutional model. The code produces no output beyond the new model file.

The fully convolutional model will convolve over arbitrary-sized input images as long as they are at least 64 × 64. The base model test set is therefore fair game, as illustrated by *kestrel_test_fcn.py*:

```
> python3 kestrel_test_fcn.py kestrel_fcn.keras
[[453    2]
 [  0 456]]
Test set accuracy: 0.9978, MCC: 0.9956
```

The results match those of the base model indicating that the fully convolutional model is working.

The fully convolutional model produces output corresponding to the size of the input, something more than a standard softmax vector. If that's the case, then how is *kestrel_test_fcn.py* producing a simple confusion matrix? Let's examine the code and the output's dimensions to understand what is happening.

The code loads the 64 × 64 test set (xtst and ytst) and the trained fully convolutional model (model) before calling predict:

```
pred = model.predict(xtst, verbose=0)
plabel = np.argmax(pred.squeeze(), axis=1)
cm,acc = ConfusionMatrix(plabel, ytst, num_classes=2)
mcc = matthews_corrcoef(ytst, plabel)
print(cm)
print('Test set accuracy: %0.4f, MCC: %0.4f' % (acc,mcc))
```

The code is nearly identical to that used in *kestrel_base.py* to test the base model, the difference is in the argument to np.argmax. It's pred.squeeze, not merely pred. Let's examine the shape of pred:

```
(Pdb) pred.shape
(911, 1, 1, 2)
```

I used Python's pdb module to break after the call to predict. There are 911 test samples, so we expect pred to be of shape 911 × 2 where each row is the softmax output corresponding to not-a-kestrel (column 0) and kestrel (column 1). Instead, there are two extra dimensions, each of size 1. As we'll learn, these are height and width dimensions that depend on the size of the input image.

The fully convolutional model is returning, for each test sample, a 3-dimensional array, (H,W,C), where C = 2 for the two classes and H and W spatial dimensions related to the size of the input image. What determines H and W for arbitrary input images will be reviewed shortly, but we know the base model expected 64 × 64 inputs, implying that the fully convolutional model built from it will operate over inputs as if it were using a 64 × 64 convolution kernel.

The base test images are 64 × 64 pixels, so there is only one convolution over the input image in this case, leading to H = 1 and W = 1. Therefore, the output of `predict` is a 1 × 1 × 2 array for each of the 911 test samples. The `squeeze` method removes dimensions of size 1 to map the output array from 911 × 1 × 1 × 2 to the expected 911 × 2.

Understanding the FCN Code

It's time to dive into the details of *kestrel_fcn.py*. We begin with the body of the `LeNet` function presented in Listing 8.1.

The listing shows a sequence of Keras layers with two marked by arrows (→). The layers on the left build the base model. Notice that there are two `Dense` layers. The arrows tell us that the fully convolutional model (what `LeNet` actually constructs) replaces these layers with new convolutional layers.

```
inp = Input(input_shape)
 _ = Conv2D(16,(3,3),padding='same')(inp)
 _ = BatchNormalization()(_)
 _ = ReLU()(_)
 _ = MaxPooling2D((2,2))(_)
 _ = Conv2D(32,(3,3),padding='same')(_)
 _ = BatchNormalization()(_)
 _ = ReLU()(_)
 _ = MaxPooling2D((2,2))(_)
 _ = Conv2D(64,(3,3),padding='same')(_)
 _ = BatchNormalization()(_)
 _ = ReLU()(_)
 _ = Flatten()(_)
 _ = Dense(128)(_)              → _ = Conv2D(128,(16,16))(_)
 _ = BatchNormalization()(_)
 _ = ReLU()(_)
 _ = Dropout(0.5)(_)
 _ = Dense(num_classes)(_)      → _ = Conv2D(num_classes,(1,1))(_)
outp = Softmax()(_)
```

Listing 8.1 Base model and layers updated to form the fully convolutional model. The `Flatten` layer is removed.

To understand the structure of the new convolutional layers, it is helpful to follow how the base model alters layer inputs and outputs. The `summary` method on a model object gives us what we need:

```
InputLayer             64 × 64 ×  3
Conv2D                 64 × 64 × 16
BatchNormalization     64 × 64 × 16
ReLU                   64 × 64 × 16
MaxPooling2D           32 × 32 × 16
Conv2D                 32 × 32 × 32
BatchNormalization     32 × 32 × 32
ReLU                   32 × 32 × 32
MaxPooling2D           16 × 16 × 32
Conv2D                 16 × 16 × 64
BatchNormalization     16 × 16 × 64
ReLU                   16 × 16 × 64
Flatten                16384
Dense                  128
BatchNormalization     128
ReLU                   128
Dropout                128
Dense                  2
Softmax                2
```

The layers read top to bottom from the input, a $64 \times 64 \times 3$ RGB image, to the softmax output, a 2-element vector. The dimensions listed indicate the shape of the tensor produced as output by that layer. The convolutional layers use zero-padding to preserve the spatial dimensions of their inputs, implying that the 2×2 max-pooling layers are altering the height and width of the layer outputs by a factor of two each time. Notice, then, that the last layer before flatten is, spatially, 16×16 and that it leads into a dense layer with 128 nodes. Recall that $16 \times 16 \times 64 = 16384$.

In other words, the base model dense layer transforms the $16 \times 16 \times 64$ input tensor into a 128-element output vector. This explains the first new convolutional layer in Listing 8.1:

```
_ = Conv2D(128, (16,16))(_)
```

We want to convolve over the now arbitrary spatial dimensions of the input tensor using a 16×16 kernel to produce 128 channel outputs. More importantly, we want the number of weights necessary to support the new convolutional layer to match those learned by the base model.

The base model learned a 16384×128 weight matrix for the 128-node dense layer. The convolutional portion of the model produces a $16 \times 16 \times 64$ output tensor, implying we can reshape the 16384×128 weights as a $16 \times 16 \times 64 \times 128$ tensor to fill in the weights of the new convolutional layer. The same logic implies mapping the 128×2 weights of the second dense layer to $1 \times 1 \times 128 \times 2$ for the 1×1 kernel convolved over the input to that layer.

The bulk of the code in *kestrel_fcn.py* copies the base model weights to the new model verbatim as those layers are unchanged, then copies the dense layer weights to the new convolutional layers after reshaping them; consider Listing 8.2.

```
model = LeNet((None,None,3), 2)
base = load_model(sys.argv[1])

for i in [1,2,5,6,9,10]:
    w = base.layers[i].get_weights()
    model.layers[i].set_weights(w)

w = base.layers[14].get_weights()
model.layers[13].set_weights(w)

w = base.layers[13].get_weights()
model.layers[12].set_weights(
                    [w[0].reshape([16,16,64,128]), w[1]])

w = base.layers[17].get_weights()
model.layers[16].set_weights([w[0].reshape([1,1,128,2]), w[1]])

model.compile(loss=keras.losses.categorical_crossentropy,
              optimizer=keras.optimizers.Adam())
model.save(sys.argv[2])
```

Listing 8.2 Filling in the weights of the fully convolutional model.

First, the fully convolutional model is defined (model). Notice that the model is told the input shape is (None,None,3). Passing None for the height and width tells Keras that the model will accept arbitrary-sized inputs. The last value of the tuple marks the inputs as having 3 channels (RGB images). The second argument is the number of classes, here 2 as we have a binary model. Recall that model is newly initialized, and that we want to overwrite the randomly assigned weights with the weights learned by the base model.

The for loop iterates over specific layers, those that are unchanged between the two models, to first read the weights into w and then copy them to the corresponding layer of the new model.

The new model lacks the Flatten layer found in the base model causing a layer offset of one between them. This explains the transfer of weights from layer 14 of the base model, a batch normalization layer, to layer 13 of the new model.

At this point, only the two new convolutional layers need to be updated by mapping the base weights of layer 13 (the 128-node dense layer) to layer 12 of the new model after reshaping to match the form of the weight expected by the new layer. The same sort of reshaping maps the second dense layer weights as well. Only the weights (w[0]) need reshaping, not the associated bias values (w[1]). Finally, we compile the new model even though we are not training it and store it on disk.

Whew! Transforming a standard CNN into a fully convolutional one requires careful consideration and multiple steps. Let's test our new model against arbitrary real-world images.

8.3.3 Real-World Examples

Let's find out if all our efforts are worthwhile. The *examples* directory contains multiple images of kestrels (*kestrel?.png* for ? in 0–9). Let's test the code in *kestrel_detection.py*, which applies the fully convolutional model to arbitrary images. The next section reviews the implementation.

The code expects the following on the command line:

```
kestrel_detection <threshold> <fcn> <image> <outdir>
    <threshold> - threshold (keep if above)
    <fcn>       - fully convolutional kestrel model (.keras)
    <image>     - image to classify
    <outdir>    - output directory
```

The model produces softmax vectors per pixel (more on that in the next section), so the user-supplied threshold is a softmax likelihood in [0, 1]. When selecting a value, remember that CNNs are often overconfident. The remaining arguments are the fully convolutional model, a .keras file, the image to classify, and a place to dump output files.

For example, let's try detecting the kestrel in the first example image:

```
> python3 kestrel_detection.py 0.998
                          kestrel_fcn.keras
                          examples/kestrel0.png
                          tmp
```

You'll first notice that the code returns in a matter of seconds, about 3 on my test machine. Fully convolutional models run fast. The output directory, *tmp*, contains:

```
background.npy
kestrel_detect.npy
kestrel_raw.npy
original_image.png
overlay_image.png
```

The PNG files are the supplied (original) image and an overlay image. The overlay image is the original image is in grayscale, with detected pixels highlighted in red with transparency. Figure 8.8 shows the detections for the first kestrel image.

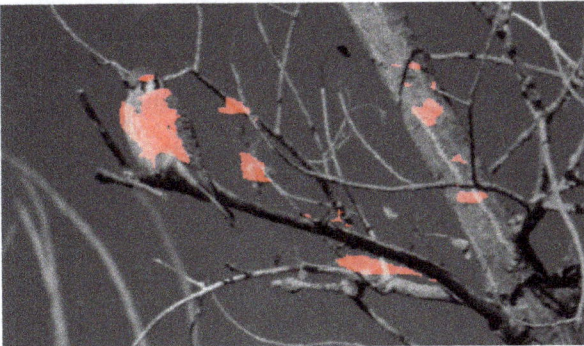

Figure 8.8 The first kestrel image and detected pixels

Figure 8.9 Additional kestrel detections

The model produces two outputs per pixel, one for the background (class 0) and one for "it's part of a kestrel" (class 1). The background and raw NumPy files contain these softmax values. The file *kestrel_detect.npy* contains softmax values above the user-supplied threshold with other regions zero.

As Figure 8.8 indicates, the fully convolutional model was a resounding success. Well, not really. Nuance is necessary. The threshold was 0.998, quite high, but a fair portion of the tree was detected along with the bird. Was the threshold too low?

Bumping the threshold to 0.9995 returns an overlay image where the central portion of the bird is detected, but about the same number of tree pixels remain scattered about. So, thresholding alone won't clean up the detected image – more about that in the discussion section at the end of the chapter.

Figure 8.9 presents results for several other example images, all with a threshold of 0.9. The kestrel was detected in each case, sometimes cleanly (top row) and sometimes less so (bottom). We claim tempered success: the model detects kestrels, if noisily.

However, should we be so immediately pleased? The training set consisted of only kestrels and background like sky and trees. What about other birds? The lower-right image in Figure 8.8 is the same flicker and kestrel image in Figure 8.6 highlighting CLIP detections with a coarse sliding window. The kestrel-only model also detects the flicker's head. Perhaps, given the training data, the model isn't so much a kestrel detector but a bird/not bird detector.

Figure 8.10 applies the detector with a threshold of 0.98 to the *ducks4* example image (yes, American Coots are not ducks; work with me here). All three birds show detections lending credence to the thought that the model hasn't learned precisely what we hoped it would learn. We'll return to this idea shortly, but first, let's review the detection code as it involves image manipulations new to us.

Figure 8.10 Applying the kestrel detector to other birds

8.3.4 Reviewing the Detection Code

Previous models required us to load an image, typically already stored as a NumPy array, then preprocess the image by scaling it to [0,1] before passing the image through a trained model the output of which was a softmax vector. The highest softmax vector value then determined the label assigned to the input image. The net effect of this process is a mapping from an input image of dimensionality $H \times W \times C$ to a d-element vector where d is the number of classes recognized by the model.

Fully convolutional models produce $h \times w \times d$ output tensors where h and w are in some (yet to be determined) relationship to H and W, the input image's spatial dimensions and d is the number of known classes. For the kestrel model, $d = 2$. With this fact in mind, let's review some of the code in *kestrel_detection.py*.

Listing 8.3 reads command line arguments, including loading the fully convolutional model and the image to classify. Notice that the image is forced to RGB space to remove any alpha channel or map grayscale images before resizing to the nearest multiple of 64, the size of the base model chips. Dividing by 255 scales image to [0, 1].

Prediction follows the usual approach: a call to predict. The single image must be reshaped to a 4-element tensor, with squeeze removing any output dimensions of

```
threshold = float(sys.argv[1])
model = load_model(sys.argv[2])
orig = np.array(Image.open(sys.argv[3]).convert("RGB"))
outdir = sys.argv[4]

h,w,c = orig.shape
sx,sy = int(64*np.round(h/64)), int(64*np.round(w/64))
img = Image.fromarray(orig).resize((sx,sy),
            resample=Image.BILINEAR)
image = np.array(img) / 255

h,w,c = image.shape
pred = model.predict(image.reshape((1,h,w,c)), verbose=0)
            .squeeze()
```

Listing 8.3 Loading, preprocessing and predicting with the fully convolutional model size 1. Therefore, the output, pred, is a 3-dimensional tensor, $h \times w \times 2$, where h and w are a function of the H and W of the resized input image.

We must align the predictions with the resized input image. The two max pooling layers, combined with 'same' convolutional padding, imply a factor of four downsampling from input to the end of the convolutional and pooling layers. The match isn't exact, but upsampling the prediction by a factor of four in each dimension will be close, and insetting the upscaled predictions with an offset determined by the difference between the resized input and the 4× upsampled output allows us to center the predictions:

```
t = zoom(pred[:,:,1], (4,4), order=1)
xx,yy = t.shape
xoff = (h - xx) // 2
yoff = (w - yy) // 2
raw = np.zeros((h,w))
raw[xoff:(xoff+xx), yoff:(yoff+yy)] = t
```

The zoom function uses bilinear interpolation on the class 1 (kestrel) prediction values to create an upsampled map which is centered in raw making raw of the same size as the resized input image. Thresholding has not yet been applied; the array contains only the raw softmax values, hence the name. The same sequence is repeated for the background softmax values.

The final set of class 1 softmax values is then a matter of resetting those below the threshold:

```
idx = np.where(raw < threshold)
kestrel = raw.copy()
kestrel[idx] = 0.0
```

The overlay image, where detections above the threshold are marked in red according to the softmax value, comes next:

```
gray =Image.fromarray((255*image).astype("uint8")).convert("L")
gray = np.array(gray)
detect = np.zeros((h,w,c), dtype="uint8")
for i in range(3): detect[:,:,i] = gray
hmap = 255*np.ones((h,w), dtype="uint8")
detect[:,:,0] = kestrel*hmap + (1-kestrel)*detect[:,:,0]
nh,nw = orig.shape[1], orig.shape[0]
detect = Image.fromarray(detect).resize((nh,nw),
        resample=Image.BILINEAR)
```

Here, image is the resized input image and orig is the image as read from disk. Therefore, the final line above resizes detect to match the original image dimensions. The remainder of the code stores output files:

```
os.system("rm -rf %s 2>/dev/null; mkdir %s" % (outdir,outdir))
np.save(outdir+"/background.npy", background)
```

```
np.save(outdir+"/kestrel_raw.npy", raw)
np.save(outdir+"/kestrel_detect.npy", kestrel)
detect.save(outdir+"/overlay_image.png")
Image.open(sys.argv[3]).convert("RGB")
          .save(outdir+"/original_image.png")
```

Building the fully convolutional kestrel model is the most complex experiment we've yet explored. The technique is only an outmoded precursor now as more advanced models, beyond our purview, have eclipsed fully convolutional approaches. However, like the fully convolutional model, and all the classic CNNs we've developed to this point, the utility of advanced networks is similarly tied to the quality of the dataset.

8.4   Discussion

This chapter explored detection, the next step beyond classification. Modern detection techniques fit into a hierarchy from "there's a bird in this image" to "there are 3 birds in this image, here's what they are and all the pixels of each." Advanced models enable such improvements, and are worth the effort to understand and implements for serious bird image analysis, to say nothing of video.

We conducted two experiments. The first applied CLIP embeddings via sliding windows as a crude attempt at detection. We already knew that CLIP embeddings enable few-shot and even zero-shot classification, meaning classifiers that require no training, or are trainable from only a few labeled instances. Running such a classifier over a larger image is the next obvious thing to do. However, we quickly realized that combinatorial explosion makes fine-grained localization impractical, to say nothing of detailed segmentation, so the exercise was pedagogical in intent. Others have pushed to combine CLIP with segmentation using more sophisticated models. Interested readers should search for CLIPSeg and related techniques.

The chapter's second experiment moved one level down in the hierarchy to build a fully convolutional model for kestrel detection. We learned that fully convolutional models are fast at inference time but subject to noise, and do not segment the object as semantic and instance segmentation models do. However, when combined with other image processing techniques, like blob detection algorithms, it is entirely reasonable to imagine a system that can indicate the presence of target species in an image, though it is likely that newer models are able to achieve similar performance with better localization. The first YOLO demo I saw was at a conference where it was running on a laptop and labeling objects in the audience at 30 frames per second.

CLIP was trained on an enormous dataset, while our little kestrel model was trained on only a few thousand images extracted from a few dozen photographs. Let it serve as another reminder that in machine learning, data is everything and that well-constructed datasets of sufficient size are essential to success.

The kestrel model clearly knows something about kestrels, especially the color patterns on the head and back, but the fact that it also picked up the Northern Flicker's head implies that it lacks the sensitivity to distinguish entirely between kestrels and other birds. Its performance on the water bird image (*ducks4.png*) supports this view. In that case, the plainer birds (American Coot and Redhead Duck) were marginally detected, but the American Wigeon was more strongly detected, especially in areas where it has a more complex color pattern similar to the kestrel.

A final lesson is that we should not assume that the model has learned as we intended it to learn. The kestrel model learned "bird-ness," but not exclusively "kestrel-ness" as we might categorize it. I'm reminded of another example from a conference presentation where an image detection model was trained to identify huskies or wolves. On the surface, it appeared to work quite well. However, when the speaker applied his "what is the model paying attention to" technique it became clear that all the wolf pictures had snow in the background while the huskie pictures did not, and that is all the model learned to detect. Models, like electricity and often people, take the path of least resistance. Larger and more representative datasets are a defense.

Birds are photogenic, endowed by evolution with a plethora of colors and plumage patterns, but beyond that, birds are noisy and produce a wide range of vocalizations. As it happens, deep learning is able to learn these patterns, as the next chapter demonstrates.

9. Classifying Audio

Birds are pleasantly vocal, hinting that at least some of their ancient dinosaur cousins were similarly expressive. In this chapter, we take a stab at classifying bird sounds. While it's possible to build deep learning systems that operate with sampled audio, a one-dimensional dataset of amplitude over time, many systems transform audio into a picture known in ornithology circles as a *sonogram*, though in other contexts it's referred to as a *spectrogram* – a two-dimensional representation of sound energy over time.

Two-dimensional data is conducive to CNNs, which regard them as yet another picture or, instead, don't regard them at all; they are merely another collection of numbers in a matrix. I think of sonograms as classic graphic equalizer displays evolving over time, the x-axis, and frequency on the y-axis. The intensity of each point represents the energy or amplitude of the frequency as a component of the audio over a short time span of a few milliseconds. If this evokes thoughts of a Fourier transform, you are on the right track.

We begin by learning how to create sonograms in code and, from there, build datasets of the same. Previous chapters leveraged the knowledge baked into CLIP embeddings. Will sonograms work in this setting? We'll learn that the answer is "no," or at least "not well" along with a likely reason why.

This leaves three options: train sonogram classification models from scratch, use transfer learning or fine-tuning. We begin with a simple, two-class transfer learning exercise, then introduce the BirdCLEF dataset, which, in true deep learning fashion, requires some preprocessing on our part.

BirdCLEF sonograms in hand, next comes training a sonogram model from scratch using the LeNet-style model from Chapter 8; see the file *kestrel_base.py*. Performance will not be stellar, but sufficient to give us some hope that transfer learning using models pretrained on ImageNet might move us to a better place. Further experiments validate this hope. Finally, a fine-tuning exercise leads us to the best performing of the chapter's models. We close with a discussion.

9.1   Sonograms

Sound evolves over time. Digitizing sound samples at discrete intervals to transform an analog voltage produced by sound impacting on a surface (i.e., a microphone) into a stream of numbers read from an analog-to-digital (A/D) converter. Digital music has been with us for decades, and most people have an intuitive understanding of this process. For us, we need only know that the samples become a long vector of numbers, the next read after a fixed interval of time from the last, implying a specific sampling rate. Fortunately, the Python libraries we need hide most of the details to let us focus on the important task of building and training models for bird audio.

It is possible to train models directly on the vector of sound samples, often decimated (literally or even more so) to produce a vector representing a few milliseconds to several seconds of sound. Classical machine learning models deal with such vectors of numbers, and even more effective are one-dimensional convolutional neural networks. The samples are in time, implying structure across the samples that such a 1-D CNN can put to good effect.

However, even more is gained from sonograms, the two-dimensional pictures of the sound energy by frequency over time. Figure 9.1 presents examples of four different bird calls: Blue Grosbeak, Great Horned Owl, Wood Duck and Northern Flicker (clockwise from the upper left). You'll find the images and the source sound files in the *examples* directory.

Sonograms are read left to right – the time axis. At each point along that axis lies a spectrum running vertically. The spectrum indicates the amount of sound energy in that frequency bin with the lowest frequencies at the bottom and the highest at the top. The higher the energy, the darker the point in the sonograms of Figure 9.1 as an inverted grayscale color table was used. Sonograms are often more colorful, but adding color buys nothing when using them as training data for neural networks.

For example, the Great Horned Owl produces a series of low frequency hoots, which appear as thin, dark, wavy lines in the upper right of Figure 9.1. The squeaks of the Wood Duck, by contrast, cover a wide range of frequencies, implying many terms of a Fourier series are needed to approximate the sound, presenting as vertical wavy lines representing the series of sinusoids summed to produce what is heard.

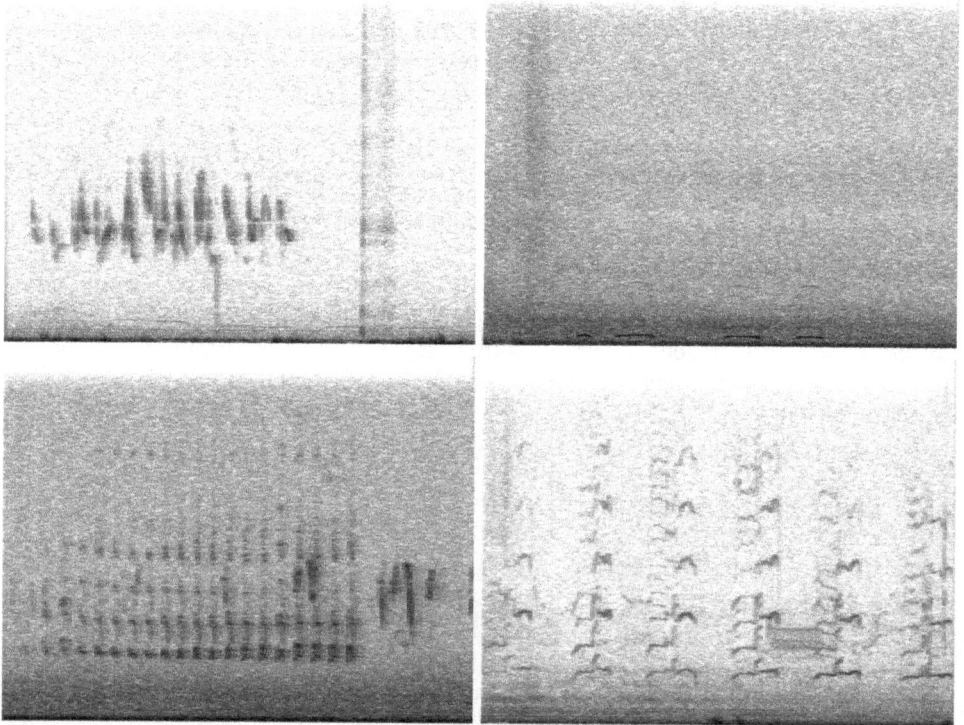

Figure 9.1 Sonograms (clockwise from upper left): Blue Grosbeak, Great Horned Owl, Wood Duck, Northern Flicker.

```
y,sr = librosa.load(wname, duration=5.0)
S = librosa.stft(y)
S_db = librosa.amplitude_to_db(np.abs(S), ref=np.max)

plt.figure(figsize=(12,9))
librosa.display.specshow(S_db, sr=sr, x_axis=None,
                          y_axis=None, cmap='gray_r')

plt.axis('off')
plt.subplots_adjust(left=0, right=1, top=1, bottom=0)
plt.savefig(sname, bbox_inches='tight', pad_inches=0)
plt.close()
```

Listing 9.1 Generating a sonogram

The sonograms of Figure 9.1 were produced by the code in *sono.py*. The bulk of the effort is buried within the librosa library which you can install with pip3:

```
> pip3 install librosa
```

The code in *sono.py* is straightforward: load the desired sound file, up to 5 seconds worth, then use librosa to generate the sonogram data plotted by Matplotlib:

Listing 9.1 contains the essentials of *sono.py*. The first code paragraph loads the sound file (wname) from the command line. The load function returns the samples (y) and the sample rate (sr) in Hertz (samples per second).

For example, loading the Great Horned Owl sound file from the *examples* directory returns sr=22050 and y a vector of 110,250 samples, all floating-point numbers in the range [− 1, 1]. Notice that the sound files are all mono, not stereo, and that load resamples to 22,050 Hz from the original 44,100 Hz in the sound file. When making recordings for conversion to sonograms, I recommend using a single channel (mono).

The next line in Listing 9.1 applies a short-time Fourier transform to move from a one-dimensional collection of sound samples in time to a two-dimensional representation of frequency and time by applying a Fourier transform to a set of overlapping windows, each about 100 milliseconds in length. The output of the Fourier transform is complex-valued.

The following line first applies np.abs to convert the complex transform output in S, already a sonogram, into a magnitude sonogram before passing that to amplitude_to_db to rescale to decibels,

$$S_{db} = 10 \log_{10}\left(\frac{|S|}{|S|_{max}}\right)$$

The second code block configures Matplotlib, along with Librosa's display module, to display the sonogram using an inverted grayscale color table before writing the sonogram to disk.

Sonograms are images. We know that CLIP embeddings transform images into 768-dimensional vectors packed with information. Therefore, it is reasonable to believe that CLIP embeddings of sonograms can help implement bird sound classifiers. Let's test this hypothesis.

9.2   A CLIP-tastrophe

This section compares CLIP embeddings of sound samples with embeddings from a reference set of sounds that we know belong to a specific collection of birds. We'll work with the following birds: Black-bellied Whistling Duck, Blue-gray Gnatcatcher, Carolina Wren, Limpkin, Northern Cardinal, Northern Mockingbird, Northern Parula, Red-shouldered Hawk, Sandhill Crane and White-eyed Vireo; see Figure 9.2.

Let's use the sound files to construct a small dataset of CLIP embeddings and then explore the embeddings to evaluate CLIP's potential in this area.

9.2.1   From Sonograms to Embeddings

The reference sound recordings come from Cornell University's Macaulay Library's collection (`search.macaulaylibrary.org`). I selected sound samples marked as available for commercial use. For each bird in the list, I downloaded example recordings and then manually split them into short clips a few seconds long. The end result was a collection of 187 sound files with between 13 and 29 per bird.

I recorded the samples near Lithia, Florida, in May of 2023 and 2024 using the Merlin app discussed in Chapter 10. Each recording was identified by Merlin as to species and clips extracted from the recordings led to between 2 to 9 examples for each species. The sound clips were transformed into sonograms and then embedded vectors by the code in *build_datasets.py*.

Please review the code in *build_datasets.py*. It's a mix of the code in Listing 9.1 and the process used in previous chapters to create CLIP embeddings from images and text (e.g., see Listing 6.2). Running the code produces multiple NumPy files included in the book's data package:

```
reference_labels        samples_labels        name_labels
reference_features      samples_features      name_features
reference_sonograms     samples_sonograms
```

Figure 9.2 Florida birds: Black-bellied Whistling Duck, Blue-gray Gnatcatcher, Carolina Wren, Limpkin, Northern Cardinal (top), Northern Mockingbird, Northern Parula, Red-shouldered Hawk, Sandhill Crane, White-eyed Vireo (bottom). (Gnatcatcher, Wren, Mockingbird, Parula and Vireo images Wikimedia Commons public domain)

The files contain sonogram images, CLIP embeddings and associated class labels identifying the bird:

```
0 Black-bellied Whistling Duck    5 Northern Mockingbird
1 Blue-gray Gnatcatcher           6 Northern Parula
2 Carolina Wren                   7 Red-shouldered Hawk
3 Limpkin                         8 Sandhill Crane
4 Northern Cardinal               9 White-eyed Vireo
```

9.2.2 Exploring the Sonogram Embeddings

Figure 9.1 makes it plain that different bird sounds produce different sonograms. Are CLIP-derived features based on sonograms useful? We have an initial dataset, so let's put it to use to learn the answer. The necessary code is in *clip_sono.py*; it's a straightforward application akin to those of Chapter 7: load the reference and sample embeddings, then, for each sample embedding, locate the *n* closest reference embeddings based on the cosine distance. The relevant portions of *clip_sono.py* are in Listing 9.2.

The main portion of the code loads the embeddings, checks the mode, and then locates the *n* closest reference embeddings for each sample. The Average function calculates the per-class average vector over the reference embeddings. Notice that using name for the mode replaces the reference embeddings with embeddings based on the bird's common name. In other words, the embeddings derived from the sample sonograms are, in that case, compared with text embeddings.

```
topn = int(sys.argv[1])
mode = sys.argv[2].lower()

samples = np.load("samples_features.npy")
labels = np.load("samples_labels.npy")
ref = np.load("reference_features.npy")
lbl = np.load("reference_labels.npy")

if (mode == "avg"):
    ref,lbl = Average(ref,lbl)
elif (mode == "name"):
    ref = np.load("name_features.npy")
    lbl = np.load("name_labels.npy")

for i in range(len(samples)):
    scores = []
    for j in range(len(ref)):
        scores.append(Cosine(ref[j], samples[i]))
    order = np.argsort(scores)[:topn]
    print("%s" % Label(labels[i]))
    for k in order:
        print("  (%0.6f) %s" % (scores[k], Label(lbl[k])))
```

Listing 9.2 Comparing reference and sample embeddings

Let's compare the samples to name embeddings returning the top-1 match:

```
> python3 clip_sono.py 1 name
```

The output consists of the sample species followed by the top-1 match and associated cosine distance. I suspect you'll notice a theme: most samples are paired with White-eyed Vireos. While this means that the output is almost entirely correct for White-eyed Vireo samples, I wouldn't read much into that. It's similar to the case where an imbalanced dataset leads to a model that is highly accurate by always selecting the most commonly occurring class.

Another observation is that the cosine distances are all, with only a few exceptions, above 0.8. Recall that orthogonal vectors have a cosine distance of 1. Therefore, name embeddings are almost orthogonal to sonogram embeddings. Back in Chapter 7, we encountered zero-shot models using CLIP embeddings reporting cosine distances between images and names all below 0.8 for the best matches, often less than 0.7. In other words, CLIP embeddings of sonograms have little association with the embeddings of common names.

CLIP embeddings are not as information-rich when based on sonograms. This is likely because CLIP was trained on image and text pairs and there are simply not enough examples out there online of sonograms labeled with a common name to let CLIP learn appropriate embeddings. Recent research supports this claim as it reveals that an exponential increase in training examples is necessary for a linear increase in embedding performance.[1]

If CLIP embeddings are not useful in this context, what about transfer learning using a model pretrained on ImageNet? We did exactly this in Chapter 6 using the code in *imagenet_features.py* and *imagenet.py*, both of which are in this chapter's directory with appropriate modifications for reference and sample sonograms.

Feature generation in Chapter 6 relied on data augmentation involving standard image manipulations like shifts and rotations. Sonograms are images, but not of a kind amenable to such alterations. So, how might we augment sonograms?

The x-axis of a sonogram represents time and the y-axis frequency with the intensity proportional to the amount of energy at that frequency. Shifting the sonogram in y, then, becomes shifting the sound in frequency, which is not what we want. We can, however, shift in the x direction as that is nothing more than altering the position of the sound in time. The Augment function in *imagenet_features.py* does just this:

```
def Augment(im):
    img = Image.fromarray(im)
    xs = np.random.randint(-20,21)
    im = np.roll(np.array(img), xs, axis=1)
    return im
```

The sonogram (im) is rolled (shifting with wrapping) between 0 and 20 pixels left or right along the x-axis.

---

1 See Udandarao, *et al*, "No 'zero-shot' without exponential data: Pretraining concept frequency determines multimodal model performance." (NeurIPS 2024).

The code in *imagenet_features.py* produces train and test sonogram embeddings based on ResNet-50 or MobileNet models pretrained on ImageNet using an augmentation factor of 10. Create the feature files by running the code twice:

```
> python3 imagenet_features.py resnet resnet
> python3 imagenet_features.py mobile mobile
```

With the first call placing ResNet-50 features in the directory *resnet* and the second MobileNet features in *mobile*. In each case, the training features are from the reference sonograms and test from the samples. For example, training a top-level MLP with 256 nodes in its hidden layer using ResNet-50 features produces:

```
> python3 imagenet.py mlp 256 resnet

MLP 256, ResNet-50 features:
0.0000  [ 0   0   0   2   0   0   0   0   0   0]  Whistling Duck
0.6667  [ 0   2   1   0   0   0   0   0   0   0]  Blue-gray Gnatcatcher
0.3333  [ 0   0   3   4   0   0   0   0   0   2]  Carolina Wren
1.0000  [ 0   0   0  11   0   0   0   0   0   0]  Limpkin
0.5000  [ 0   0   0   1   1   0   0   0   0   0]  Northern Cardinal
0.2000  [ 0   0   2   1   0   1   1   0   0   0]  Northern Mockingbird
0.5000  [ 0   0   0   1   0   0   1   0   0   0]  Northern Parula
0.1250  [ 0   0   0   7   0   0   0   1   0   0]  Red-shouldered Hawk
0.4000  [ 0   0   0   3   0   0   0   0   2   0]  Sandhill Crane
0.4286  [ 0   0   0   4   0   0   0   0   0   3]  White-eyed Vireo
Test set accuracy: 0.4630, MCC: 0.4128
```

The output shows a confusion matrix for the samples with the per-class accuracy first followed by model class assignment counts and the class name. The number of test samples is small, but trends are evident.

My run produced a top-level model with an overall accuracy of 46 percent and per-class accuracies ranging from 0 to 100 percent. However, the number of samples in the test set makes it difficult to believe the results are particularly robust. The largest number of samples are from Limpkins, which are perfectly classified, but running your eye down the confusion matrix uncovers that the model is quite fond of Limpkin sonograms, perhaps because of the Limpkin's unique sound; play *limpkin.wav* in the *examples* directory.

Feel free to experiment with MobileNet features and with different-sized MLPs or random forests. I expect you'll come to the same conclusions I did: MLPs outperform random forests and ResNet-50 features are more useful than MobileNet features.

This section informed us of the inadequacy of CLIP features for sonograms and hinted that features derived from pretrained models might be more applicable. The next section builds on this intuition to learn if transfer learning is capable of distinguishing between two closely related species.

9.3   A Transfer Learning Exercise

Canada Geese are ubiquitous throughout Canada and the continental United States. Cackling Geese are visually quite similar, so much so that many people don't realize the two are separate species as they tend to flock together. Cackling Geese have a more restricted range compared to Canada Geese, but the two overlap significantly in central Colorado, where I live. In the winter, it's often true that most of the geese are Cackling and not Canada.

The two species sound much the same as well. The Merlin app (see Chapter 10) identifies both species when listening to the noise of a large, mixed group. In this section, we use transfer learning to make a narrowly focused classifier attempting to do the same.

The following files contain sonograms of each species generated from commercially available sounds in the Macaulay library collection:

```
goose_xtest.npy
goose_xtrain.npy
goose_ytest.npy
goose_ytrain.npy
```

I split the recordings into train and test sets using the code in *build_goose.py*, which uses librosa to extract 2-second clips from the set of longer sound files.

The sonograms are already resized to 224×224 pixels and made RGB to meet the input requirements of pretrained ResNet-50 and MobileNet models. There are 204 training samples, 100 Canada (class 0) and 102 Cackling (class 1). The test set is evenly split at 26 samples each.

The dataset is small. Training a CNN from scratch is perhaps unrealistic (prove me wrong), so instead, we again turn to transfer learning using large architectures pretrained on ImageNet. The ImageNet dataset does not contain sonograms, but it is likely to have learned at least low-level feature detectors in its first layers that are (hopefully) capable of producing sufficiently distinct output embeddings to enable a top-level model to discern between Canada and Cackling geese.

First, run *goose_features.py* twice to create the expected ResNet and MobileNet embeddings:

```
> python3 goose_features.py resnet resnet_goose
> python3 goose_features.py mobile mobile_goose
```

Then, use *goose.py* to train and test random forest or MLP top-level models. For example,

```
> python3 goose.py mlp 256 resnet goose_resnet_mlp_256
MLP 256, ResNet-50 features:
[[24  2]
 [ 1 25]]
Test set accuracy: 0.9423, MCC: 0.8853
```

The results are encouraging. The model is 94 percent accurate on the test set, clearly distinguishing between the two species. As previously noted, other combinations of top-level model and source features are not quite as good but are still potentially useful. For example, a small random forest using MobileNet features produces:

```
> python3 -W ignore goose.py rf 100 mobile goose_mobile_rf_100
RF 100, MobileNet features:
[[23  3]
 [ 4 22]]
Test set accuracy: 0.8654, MCC: 0.7313
```

The random forest results are still in the useful range. This exercise is a reminder of the utility often found in restricting the set of known inputs to a model. The classifier is narrowly focused, and because of this performs well even with limited data under the assumption that the input is already known to be either a Canada or Cackling goose.

What we require is a larger sonogram dataset. Fortunately, one is available, though some effort is necessary to get it ready for our use.

9.4   Preparing the BirdCLEF Dataset

BirdCLEF is a Kaggle competition hosted by the Cornell Lab of Ornithology. It contains audio recordings of dozens of different bird species and invites competitors to submit their models with those achieving the highest overall accuracy winning prizes. The BirdCLEF dataset has been cast as sonograms. To use it, we must download the dataset from Kaggle, which requires a free account. As previously mentioned, signing up with Kaggle is worthwhile to gain access to the many datasets it provides (and competitions, if interested). Download the sonograms by clicking the "Download" button found here:

www.kaggle.com/datasets/richolson/birdclef-2024-mel-spectrograms

I used the download as a .zip file option and expanded in the chapter's directory to produce a *BirdCLEF* folder containing *train_images*, which itself contains 182 directories of bird images. The directory names match how `ebird.org` references the bird in question. I mapped 180 of these directories to common bird names and placed them in *common_names.txt*. This file is used by *build_birdclef.py* to generate the dataset, as we'll use it here. Run the file to produce the output datasets prefixed by *birdclef*:

```
> python3 build_birdclef.py
```

The code separates *common_names.txt* into directory and common name, then builds train and test NumPy arrays using a 70 to 30 split. The training sonograms are augmented in time by a factor of 3x. Please review the code; it's quite similar to several dataset-building scripts we've previously employed.

The resulting dataset consists of 180 classes with 30,075 training samples and 4419 test samples. Let's use this dataset to train a model from scratch before exploring transfer learning and fine-tuning.

9.5   Training BirdCLEF from Scratch

The BirdCLEF dataset contains a minimum of 12 (Oriental Darter, Nilgiri Wood-Pigeon) to a maximum of 243 (Malabar Whistling-Thrush) examples per species. The median is 190. Is this enough to successfully train a CNN from scratch? To find out, consider the code in *lenet.py*, which is a close cousin to the LeNet-5 model we used in earlier chapters with an updated set of convolutional filters (16, 32, 64) and 128 nodes in the dense layer at the top. It follows our usual approach, so I will refrain from detailing the code here. However, the training set in image form is rather extensive: $30,075 \times 224 \times 224 \times 3 = 4,527,129,600$ bytes, so the code allows for specifying a smaller subset before training. Additionally, the normalized images ([0, 1]) are stored as 16-bit floats:

```
xtrn = xtrn.astype('float16') / 255
xtst = xtst.astype('float16') / 255
```

Reduced-precision floating-point values, or even scaled integer values, are often used in deep learning to minimize memory use. Neural networks are surprisingly robust, in general, to the loss of precision. Notice, however, that storing the normalized images as 16-bit floats does not carry forward to actual training; all calculations are performed at 32-bit precision in hardware.

Let's give *lenet.py* a go to establish a baseline level of performance for the BirdCLEF dataset. The command line expects the minibatch size, number of training epochs, output directory and, optionally, the number of training samples to retain. For example,

```
> python3 lenet.py 32 9 lenet 14000
```

uses a minibatch of 32, trains for 9 epochs, and retains 14,000 of the 30,075 available training samples. Why 14,000? Because that is the maximum number my 32 GB RAM test machine would accommodate.

When training was complete, *lenet.py* reported the results, including the overall accuracy and Matthews correlation:

```
Test set accuracy: 0.2399, MCC: 0.2357
```

With 180 classes, we would expect an accuracy closer to 0.6 percent if the model were merely guessing class labels. The LeNet-style model's accuracy of 24.0 percent indicates that it did successfully learn something about the appearance of sonograms. The top-10 best classified species, courtesy of *lenet_analysis.py*, were:

```
0.7576   Rosy Starling
0.7143   Jungle Nightjar
0.6765   Asian Emerald Dove
0.6364   Common Hawk-Cuckoo
0.6000   Gray-bellied Cuckoo
0.6000   Greater Flameback
0.5882   Black-and-orange Flycatcher
0.5806   Eurasian Hoopoe
0.5667   Asian Koel
0.5625   Brown Boobook
```

Overall, the mean per-class accuracy was 21.4 percent with a median of 16.7 percent. Notice that the per-class mean can be less than the overall accuracy because the former treats each class equally while the latter is implicitly weighted by the number of test samples in each class.

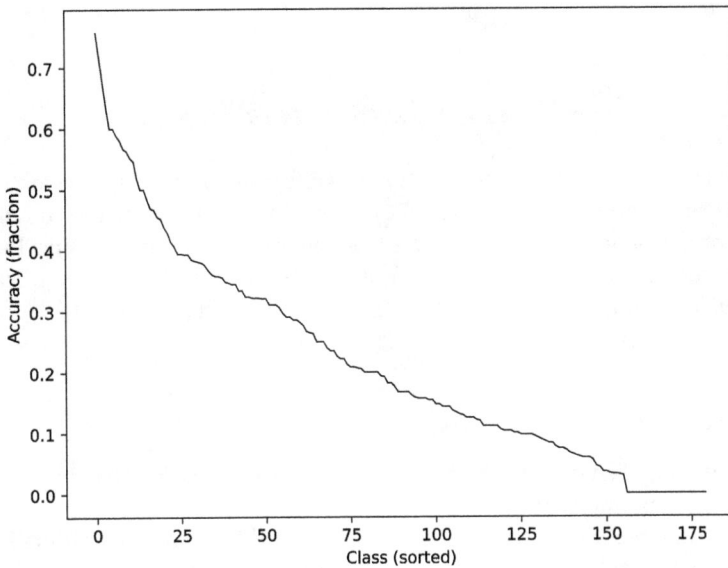

Figure 9.3 Per-class BirdCLEF accuracies sorted from highest to lowest for the LeNet-style model.

Figure 9.3 shows the per-class accuracies sorted from highest to lowest. Only 13 species were classified with an accuracy of above 50 percent with 24 species failing completely (accuracy of zero). The LeNet-style model works for a handful of species but struggles significantly with many of the others.

We have a baseline level of performance. Now, let's try to do better with transfer learning as we did in Chapter 6.

9.6 BirdCLEF Transfer Learning

The file *birdclef_features.py*, which is essentially the same as *imagenet_features.py* from Chapter 6, produces ResNet-50 or MobileNet output features using the BirdCLEF sonograms. Run it twice:

```
> python3 birdclef_features.py mobile mobile_birdclef
> python3 birdclef_features.py resnet resnet_birdclef
```

The file *birdclef.py*, a nearly complete clone of *imagenet.py* from Chapter 6, trains top-level random forests or MLPs using the selected features. For example, the following trains an MLP using 256 nodes in the first of its two hidden layers:

```
> python3 -W ignore birdclef.py mlp 256 resnet tmp
MLP 256, ResNet-50 features:
[[16   0   1 ...   0   0   0]
 [ 0   6   0 ...   0   0   0]
 [ 0   0  10 ...   0   0   0]
 . . .
```

```
[ 0   0   1 ... 10   0   0]
[ 0   0   0 ...  0  11   0]
[ 0   0   0 ...  0   0  16]])
Test set accuracy: 0.3480, MCC: 0.3439
```

Transfer learning with ResNet-50 features results in an improvement over the baseline result, 35 percent overall accuracy versus 24 percent, but we shouldn't be too satisfied with this level of performance. True, it's over 60 times better than random guessing, but I suspect no one would want to use this model for any real-world task. Feel free to experiment with MobileNet features and random forests. I suspect you'll find that MLPs with ResNet features perform best.

9.7   BirdCLEF Fine-Tuning

We know that ImageNet does not contain sonograms, so it stands to reason that we might gain, in this case, by fine-tuning.

Let's fine-tune MobileNet (because ResNet-50 is too big) using the training sonograms, then use the fine-tuned MobileNet model to produce new embeddings for downstream models. Along the way, we'll test the fine-tuned MobileNet on the held-out test samples. Here are the necessary steps:

1.   Load the train and test sonograms (no embedding vectors yet)
2.   Load MobileNetV3Large pretrained on ImageNet
3.   Freeze the first 80 base model layers
4.   Add a simple classification head so we can fine-tune
5.   Train the new model for a user-specified number of epochs
6.   Test the fine-tuned MobileNet model
7.   Pass both train and test sonograms through the model
8.   Train and test top-level models on the embeddings

I'll walk through the steps as this is our first fine-tuning exercise. Most of the code is in *birdclef_fine-tune.py* with the top-level model code in *fine-tune.py*. As always, please review these files before continuing.

9.7.1   Fine-Tuning MobileNet

Run the fine-tuning code first to learn how well it performs with sonograms and to produce the embeddings expected in the next section. Then, continue to the code review.

Running the Code

The code in *birdclef_fine-tune.py* accomplishes the first seven steps above.

I fine-tuned for 10 epochs using a minibatch of 32 samples. Fine-tuning typically assumes the model to be close (in some sense) to what we desire, hence the name "fine-tuning." In those cases, conventional wisdom says to train for only a few epochs with an optimizer set to take tiny steps, i.e., using a small learning rate.

Our case, however, differs somewhat. First, we know that the pretrained model is not "close" to what we need because it was trained on images, not sonograms. Lower-level layers are likely general enough to be helpful as they are, but middle and higher layers are not well-conditioned for the task. Second, we are not intending to use the fine-tuned

MobileNet downstream. We certainly could, but I'm thinking here of comparing with our previous transfer learning results and of insisting on the smaller MLP model for the downstream task because it is lightweight in terms of parameters. Therefore, it seems reasonable to fine-tune for longer and with an optimizer set to use the default learning rate.

The file *birdclef_fine-tune.py* expects three command line parameters: the minibatch size, the number of epochs and an output directory for the new sonogram embeddings and the confusion matrix:

```
> python3 birdclef_fine-tune.py 32 10 mobile_fine-tune_32_10
```

My training session took 2 hours. The output is in the file *mobile_fine-tune_results.txt* with a test set accuracy of 0.4933, meaning fine-tuning on the sonograms improved over the 24 percent baseline by nearly a factor of two and by almost 15 percentage points compared to transfer learning without fine-tuning.

The fine-tuned embeddings are in the directory *mobile_fine-tune_32_10*. The final step trains top-level models, but before we do, let's understand the code that produced them.

A Code Walkthrough

Fine-tuning MobileNet begins by loading the sonograms as images followed by defining the pretrained MobileNet model:

```
input_shape = (224, 224, 3)
base = MobileNetV3Large(input_shape=input_shape,
                        weights='imagenet',
                        include_top=False,
                        pooling='avg')
```

The sonograms are already 224 × 224 RGB images, as indicated by `input_shape`. The object `base` is instantiated by telling Keras what input shape to use, that we want to use ImageNet weights, that there should be no top-level classification head, and that the output should use global average pooling to return a vector for each input sonogram. This is the base model to which we'll add a classification head tailored to the number of BirdCLEF classes (180). Setting `include_top` to `True` assumes 1,000 output classes – the number of classes in ImageNet.

We do not want to fine-tune every layer in the pretrained `base` model, so next, we freeze the first 80 layers and mark the remaining top-level layers as available for training:

```
N = 80
for layer in base.layers[:N]:
    layer.trainable = False
for layer in base.layers[N:]:
    layer.trainable = True
```

There's nothing magical about $N = 80$ other than it covers roughly the first third of the 263 layers in MobileNetV3Large when `include_top` is `False`. Feel free to experiment with other cutoffs. The critical point is that low-level layers are frozen because we assume them already amenable to the task, but higher-level layers are adapted to ImageNet in particular and must be retrained to work well with BirdCLEF sonograms.

We're now ready to add the task-specific classification head:

```
num_classes = 180
_ = Dense(256)(base.output)
_ = ReLU()(_)
_ = Dropout(0.5)(_)
_ = Dense(num_classes)(_)
outp = Softmax()(_)

model = Model(inputs=base.input, outputs=outp)
model.compile(optimizer=Adam(),
              loss='sparse_categorical_crossentropy',
              metrics=['accuracy'])
```

The base model output, already a vector because we specified global average pooling, is linked to a new top-level dense layer with 256 nodes followed by a ReLU, dropout and a final dense layer with softmax to produce the required 180-element output vector, the maximum value of which indicates the model's selected species.

Notice that the input to model is the input to base but that the output of model is the output of the newly appended softmax layer. This trick of building a new model linked to a base model or a desired output layer is commonly used in Keras.

Fine-tuning requires compiling the new model. As usual, we use Adam as the optimizer and request accuracy metrics during training. Specifying the loss function as sparse categorical cross-entropy is new. Previously, we converted dataset integer labels to one-hot vectors, then called compile with categorical cross-entropy as the loss function. Sparse categorical cross-entropy is the same but uses the integer labels as they are; no conversion to one-hot vectors is required.

We're now ready to fine-tune the MobileNet model:

```
x_train = preprocess_input(x_train)
x_test = preprocess_input(x_test)

model.fit(x_train, ytrain,
          validation_split=0.2,
          epochs=epochs, batch_size=mb,
          verbose=1)
```

The sonograms are passed through preprocess, a function provided by the Keras applications module, specific to each model architecture, to adjust inputs to the form expected by the base model. Training itself runs for the user-specified number of epochs using the requested minibatch size. The validation_split keyword selects a portion of the training data for validation – the accuracy reported during the training process. As always, final testing uses the x_test samples:

```
prob = model.predict(x_test, verbose=0)
plabel = np.argmax(prob, axis=1)
cm,acc = ConfusionMatrix(plabel,ytest, num_classes=num_classes)
print("Test set accuracy = %0.5f" % acc)
```

The reported test set accuracy is for the MobileNet itself using the test set sonograms as images. We want embedding vectors for downstream models, so we use the base network to produce them knowing that the unfrozen weights of the base network have been adjusted to (hopefully) better match sonograms as inputs. The embedding vectors are first found, then normalized to [0, 1]:

```
xtrn = base.predict(x_train, verbose=0)
xtst = base.predict(x_test, verbose=0)

xtrn = (xtrn - xtrn.min()) / (xtrn.max() - xtrn.min())
xtst = (xtst - xtst.min()) / (xtst.max() - xtst.min())
```

The remainder of the code dumps the embedding vectors to disk along with the confusion matrix representing the fine-tuned MobileNet's performance on the test set.

9.7.2 Using the Fine-Tuned Embeddings

The code in *fine-tune.py* trains top-level models, RF or MLP, using the embeddings generated by the fine-tuned MobileNet model. Computationally, it is identical to *birdclef.py* differing only in the embeddings used. For example,

```
> python3 fine-tune.py mlp 512 tmp
```

trains a top-level MLP with 512 nodes in the first hidden layer (256 in the second), dumping results in *tmp*. My run produced:

```
Test set accuracy: 0.5472, MCC: 0.5445
```

Fine-tuning MobileNet paid off in the end. The top-level MLP was nearly 55 percent accurate over all 180 classes, a far cry above the first LeNet-style model trained from scratch and even significantly improved over the fine-tuned MobileNet model's 49 percent accuracy.

I'm happy with this result, but to put things into perspective, an overall accuracy of 55 percent places 848-th on the BirdCLEF 2024 challenge leaderboard. The winning model managed an accuracy of a smidgen over 69 percent, so while our result isn't a winner, it's not exceedingly distant from what is possible.

The file *fine-tune_analysis.py* repeats the per-class accuracy analysis of the LeNet-style model we first trained. Running it, assuming *lenet_analysis.py* has been run first, produces this output and a plot comparing the per-class accuracies:

```
Overall accuracy: 0.5472

0.9062  Brown Boobook
0.9032  Great Eared-Nightjar
0.8667  Little Swift
0.8438  Great Hornbill
0.8276  Little Spiderhunter
0.8235  Puff-throated Babbler
0.8235  Asian Emerald Dove
```

```
0.8182   Coppersmith Barbet
0.8182   Rosy Starling
0.8065   Eurasian Hoopoe

Mean accuracy 0.5219, median 0.5455

103 species had an accuracy >0.5
2 species had an accuracy of 0
```

These results are significant improvements over the LeNet-style results from the previous section. Here, 103 species were classified above 50 percent accuracy and only 2 were complete failures. Compare this with only 13 species above 50 percent accuracy and 24 failures for the LeNet-style model.

The Asian Emerald Dove, Brown Boobook, Eurasian Hoopoe and Rosy Starling are shared in common between the two models, hinting that these birds are particularly easy to detect. However, the rate of detection by the LeNet-style model for these birds was always less than 80 percent.

Figure 9.4 presents the per-class accuracies from highest to lowest across all species along with the same data from Figure 9.3 for the LeNet-style model (dashed line). It's expected that both curves decrease across the classes, and it's important to remember that the x-axis class ordering differs between the two curves. The takeaway is that the fine-tuned embeddings lead to a large increase in the number of species reasonably detected compared to the from-scratch model.

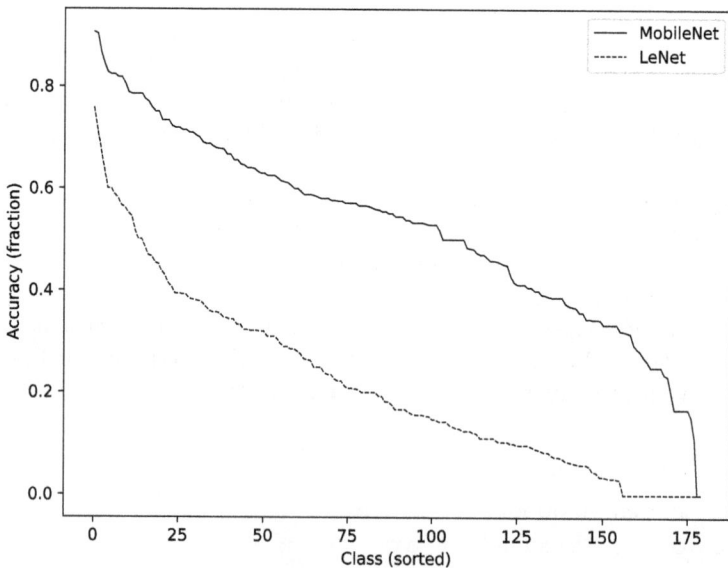

Figure 9.4 Per-class BirdCLEF accuracies for the MLP 512 model (solid) and the LeNet-style model trained from scratch on sonograms (dashed).

9.8   Discussion

This chapter sought to classify bird sounds using sonograms. The approach is reasonable, but not as easily accomplished as with images and powerful CLIP embeddings. We learned that simple models produce meager results, while transfer learning was not as helpful as we might have initially anticipated. CLIP embeddings failed outright.

We did, however, find success with fine-tuning a larger, pretrained model. It's not difficult to imagine that a more careful and tailored application of fine-tuning might improve results even more.

The BirdCLEF models try to be generalist and successful against 180 different species. The Canada and Cackling Goose exercise demonstrated that sometimes, being focused is better. As mentioned, this is a general rule in deep learning. The more the world (meaning a set of possible inputs to a model) is controlled, the better the models will be at classifying it.

The next chapter explores open-source birding tools.

10. Open Source Birding with AI

The AI revolution has already led to the creation of many AI-related birding tools. In this chapter, we explore three freely available for smartphones, the web, and desktop environments: Merlin, eBird and BirdNET Analyzer. All are products of the Cornell Lab of Ornithology at Cornell University in collaboration with Chemnitz University of Technology in Germany (BirdNET).

10.1 Merlin

The Merlin app runs on Android and iOS. It uses AI interpretation of pictures and audio recorded in the app (audio import is also supported). Further, users are able to create and link the application to their eBird accounts (https://ebird.org/), thereby providing a place for holding life lists and enabling participation in community-level birding. An eBird account is recommended.

The application's home screen, audio recording playback, and image ID screens are in Figure 10.1. I'm using an Android phone. The bird of the day is tailored to your current location. The screen capture was made in mid-March in central Colorado, hence the Red-breasted Merganser.

The audio screen shows a scrolling sonogram and detected species highlighted when that bird's call is playing. It's possible to export the recording to transfer it to a computer, as well as import recordings from other sources.

Image identification is tailored to the quality of most phones. However, you can import images from other sources, like a camera, and Merlin will identify the bird for you. New species are added to your eBird life list. In Figure 10.1, I imported a camera image of a Merlin, which the app correctly identifies. Not shown is the screen where the user selects the image's location and date. The Merlin picture is mine and was taken in Westminster, Colorado, on the date indicated; however, the app doesn't know if a photo is one the user actually took; birding is an honest person's activity, so trust is required.

Merlin is configured by downloading regional data. I use the dataset for North America, which means the app might not identify birds from other regions. Identification is aware of the image's provenance, which is helpful in cases where similar species not likely to be found in the user's current geographic region are excluded from consideration. As we'll learn later in the chapter, this paring of possibilities is a double-edged sword. Merlin uses eBird observation data to further increase the likelihood of a correct classification.

The Merlin model is not currently available outside of the application, though it's possible that might change in time (Grant van Horn, personal communication). If the

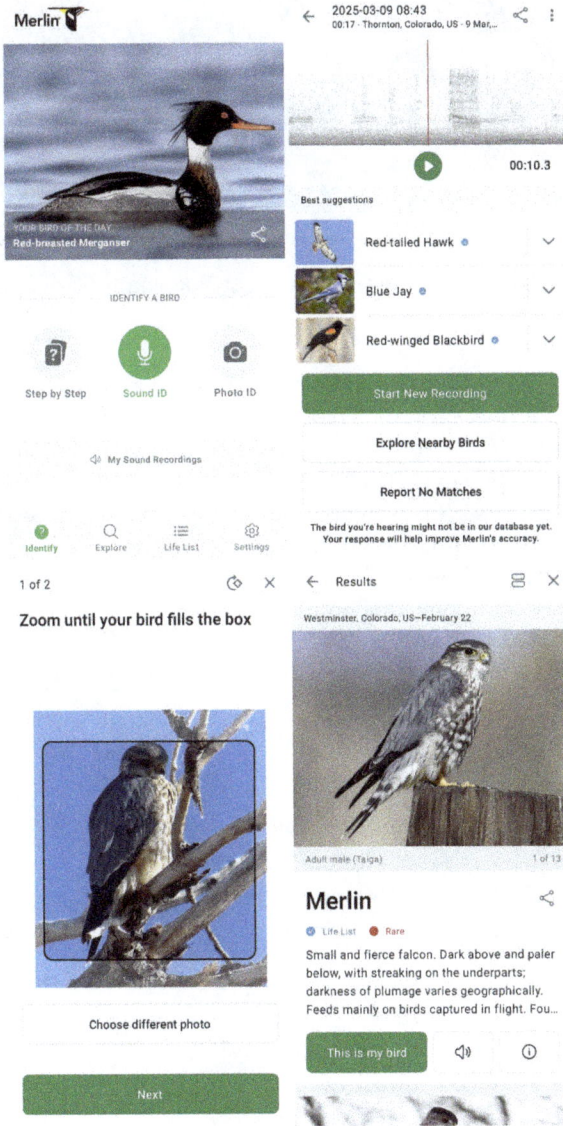

Figure 10.1 Merlin app screens: home and audio (top), image ID query and result (bottom).

model isn't available when you read this, and you are a professional, asking about the status seems warranted. Do bear in mind, however, that our CLIP embedding model using cosine distance and common bird name embeddings was highly accurate on its own, even more so when using a few-shot approach to train a top-level MLP.

The rapid increase in birding by the general public is mainly due to the Merlin app (and the global pandemic). Let's explore the app's classification abilities and compare them to CLIP-based models from Chapter 7.

The chapter's *examples* directory contains three cell phone images of birds: Great Blue Heron, Black-crowned Night Heron, and a Lark Sparrow. The images are far from sterling, but of the kind Merlin is likely to be asked to classify; see Figure 10.2.

Figure 10.2 Low-resolution cell phone images: Great Blue Heron, Black-crowned Night Heron, Lark Sparrow

I asked Merlin to identify each of the images. It was correct in every instance. That is, as long as I used the proper geographic location. I told Merlin that the bird images were taken in Scotland and not Colorado; it failed. It offered no suggestions for the Lark Sparrow and Night Heron. The app suggested a Gray Heron as a candidate identification for the Great Blue Heron. Of course, this is an unfair test and a clue about the power of including geographic information in the classification process.

How do the models of Chapter 7 fare? Using cosine distance alone compared to the text embedding of the common name produces:

```
> python3 clip_image_generic.py 3 avg great_heron__poor.jpg
  (0.087158)  Great Blue Heron
  (0.143025)  Green Heron
  (0.145523)  Tricolored Heron

> python3 clip_image_generic.py 3 avg night_heron__poor.jpg
  (0.118808)  Black-crowned Night-Heron
  (0.131133)  Yellow-crowned Night-Heron
  (0.151935)  Great Blue Heron

> python3 clip_image_generic.py 8 avg lark_sparrow__poor.jpg
  (0.154090)  Lincoln's Sparrow
  (0.156497)  Louisiana Waterthrush
  (0.159747)  Vesper Sparrow
  (0.163083)  Clay-colored Sparrow
  (0.167268)  American Pipit
  (0.170541)  Savannah Sparrow
  (0.170738)  Northern Waterthrush
  (0.171204)  Lark Sparrow
```

Recall that a smaller cosine distance is better. Both herons were correctly identified as the top-1 result. The Lark Sparrow was also present, but only at the bottom of the top-8. Applying the few-shot MLP classifier from the same chapter returned:

```
> python3 mlp_image_generic.py mlp.pkl 3 great_heron__poor.jpg
  (0.99987)  Great Blue Heron
  (0.00009)  Tricolored Heron
  (0.00002)  Cooper's Hawk
```

```
> python3 mlp_image_generic.py mlp.pkl 3 night_heron__poor.jpg
  (0.99994)   Black-crowned Night-Heron
  (0.00006)   Yellow-crowned Night-Heron
  (0.00000)   Cattle Egret

> python3 mlp_image_generic.py mlp.pkl 3 lark_sparrow__poor.jpg
  (0.37240)   Vesper Sparrow
  (0.26155)   Lincoln's Sparrow
  (0.13970)   Lark Sparrow
```

All three birds are within the top-3, with the herons the top-1 result by a wide margin. The MLP classifier returns softmax values, indicating the model was quite confident in its top response for the herons. The sparrow was at the bottom of the top-3, but the low softmax values hint that the model was quite unsure.

Let's push the envelope. I rescaled the example images' longest dimension from approximately 1000 to 200 pixels and then passed those images to the MLP classifier. The herons were still selected as top-1 with softmax values above 0.99 each. The Lark Sparrow fell from 3rd place to 5th with a softmax of 0.03, still within the typical top-5 result reported in the literature. Merlin also correctly identified the herons but offered no suggestions for the sparrow.

A final stress test reduced the resolution to a mere 100 pixels along the largest dimension. The CLIP-based MLP still correctly identifies the herons as top-1, but the Lark Sparrow falls to 15th place (out of a possible 404 species). Merlin fails to identify any of the birds.

If spoofing geographic location is an unfair test, then reducing image resolution to ridiculous extremes is unconscionable. The remarkable result isn't that Merlin, designed for modern cell phone camera resolutions, failed but that the few-shot, CLIP-based MLP could still identify two of the three birds correctly, this time with softmax values above 0.98.

Merlin's audio recording, at least with my particular Android phone, isn't as sensitive as I would like it to be, but I suspect the phone more than the app. I often hear birds that fail to register. One nice aspect of recording is the ability to export the audio files (.wav format). Bulk export is possible with other applications, such as scp or sftp tools, or by simply sharing the files via email. Once exported, it is a straighforward exercise to process the recording with standard audio tools like Audacity (open source). Note that the recordings are floating-point and likely scaled to [− 1, 1] so that they are initially too faint to see or hear in Audacity. Applying normalization fixes the issue. If the recordings are scaled that way, we now know enough about AI to understand why: neural networks like inputs in that range.

Merlin is a tool all birders, professional or otherwise, should have ready access to. It works brilliantly, integrates with online sites like eBird, and is well-crafted and implemented. I use it daily. Kudos to the app's development team!

10.2   eBird

eBird (https://ebird.org/), like Merlin, is a product of the Cornell Lab of Ornithology. Its primary purpose is to serve as a hub and data repository for both birders and professional ornithologists.

Birders benefit from the connection to Merlin and as a place to store life lists, submit observation reports, discover birding hotspots and access real-time bird distribution data.

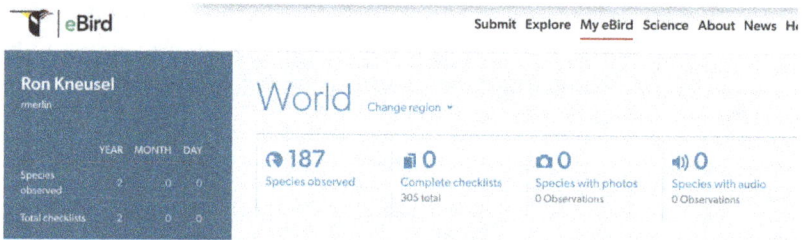

Figure 10.3 A portion of the author's "My eBird" page

An account is recommended and is linked to the Merlin app. Figure 10.3 It shows a portion of my account's "My eBird" page, humble as it is. I tend to rely on Merlin for identification and to track my life list, but I haven't yet increased my involvement with eBird to take fuller advantage of what it offers. For example, the Bird Academy portion of eBird offers an entire catalog of birding courses at reasonable prices. The "eBird Essentials" course is free and is recommended as a good place to start.

Professionals benefit from eBird's vast data collection regarding distribution, migration and population trends. Birder observations from all over the world allow researchers to interpret bird patterns in ways simply impossible in the past. Conservation efforts benefit from eBird's datasets and will only become more critical in the future due to our rapidly changing climate. The data eBird collects and disseminates means literal life or death for many bird species.

10.3 BirdNET

Merlin can identify bird sounds, but for large collections of recorded audio, BirdNET is the way to go. BirdNET runs on a laptop or desktop computer, with or without a GPU, and classifies individual .wav files or entire directories at once. The main BirdNET website also provides a place to upload a sound file; see `https://birdnet.cornell.edu/`.

Let's install BirdNET and then experiment with it to understand some of its functionality. First, install BirdNET in its own Python environment. My steps were:

```
> python3 -m venv birdnet
> source birdnet/bin/activate
(birdnet) > git clone https://github.com/kahst/
                        BirdNET-Analyzer.git

(birdnet) > cd BirdNET-Analyzer
(birdnet) > pip3 install .
(birdnet) > pip3 install keras-tuner
```

The `venv` module creates the `birdnet` environment, which is then activated. Enter `deactivate` to exit the environment.

The following line uses `git` to install BirdNET. Alternatively, you can download the .zip file directly from the GitHub site. Either approach leads to a directory named *BirdNET–Analyzer*. Configuring BirdNET means changing to that directory and running `pip3` to parse the included *pyproject.toml* file. A suite of libraries will be installed. I found that the keras-tuner library was necessary but not installed, hence installing it manually. If all goes well, BirdNET will be ready to use. I recommend copying the *examples* directory to the BirdNET directory along with the *analyze* scripts. Doing so makes it easier to run the experiments.

I included three scripts to simplify running BirdNet: *analyze_co*, *analyze_fl* and *analyze_all*. The first sets the latitude and longitude to Westminster, Colorado, the second to Lithia, Florida and the last excludes latitude and longitude to consider all species BirdNET is aware of. For example, *analyze_co* is:

```
#!/bin/sh
python3 -m birdnet_analyzer.analyze
       --lat 39.9264429 --lon -104.9534609 -o tmp $1
```

The first line allows the script to run from the command line directly. The second line (all one line in the file) is the call to BirdNET. The meaning of the --lat and --lon keywords are clear. The script outputs results to a directory named *tmp*. Finally, the $1 represents the argument to the script, either a single audio file or the name of a directory containing multiple audio files. The Florida script is the same but sets latitude and longitude appropriately. The "all" script excludes --lat and --lon.

The *examples* directory contains *red-tail_mimic.wav*, a recording of a Blue Jay mimicking a Red-tailed Hawk. Let's pass this file through BirdNET. Will it detect the deception?

```
(birdnet) > ./analyze_co examples/red-tail_mimic.wav
--snip--
(birdnet) > cat tmp/red-tail_mimic.BirdNET.selection.table.txt
1   Spectrogram    0.0   3.0     Red-tailed Hawk rethaw    0.8916
2   Spectrogram    3.0   6.0     Blue Jay         blujay    0.4889
3   Spectrogram    6.0   9.0     Red-tailed Hawk rethaw    0.9842
4   Spectrogram    9.0   12.0    Red-tailed Hawk rethaw    0.9512
```

BirdNET creates output files in *tmp*, as requested by the script. Each input file or directory of files creates a similarly-named text file in *tmp*. Therefore, examine the text file to understand BirdNET's evaluation.

The BirdNET-Analyzer evaluates audio files in 3-second disjoint increments by default. Four detections were found; three were identified as Red-tailed Hawk and one as Blue Jay. Note that the listing excludes potentially many lines of warning text from TensorFlow, especially if there is no GPU in the system (--snip--). Further, actual text file output includes other bits of information per detection, which are excluded here to focus on the most important results. I performed the experiments on an old desktop system with a quad-core i5 processor and 8 GB of memory to demonstrate that BirdNET should run almost everywhere.

The final number in each detection is a confidence, likely based on a softmax output value. The model was quite confident that it was hearing a Red-tailed Hawk, a testament to the Blue Jay's abilities as a mimic. The second detection is labeled Blue Jay, but it was a second Blue Jay in the distance (I know because I made the recording and was watching the mimic). I doubt a human would know that the hawk call was fake, so we cannot fault the model for doing the same.

The Merlin app benefits from geographic location information. BirdNET is the same, but this is sometimes a hindrance. For example, the file *swamp_sparrow.wav* contains a recording of a Swamp Sparrow I made in Colorado. The range map from All About Birds indicates that the bird is found in Florida but marked as "scarce" in Colorado. If we pass the recording to BirdNET using *analyze_co*, we receive no detection, which BirdNET denotes as `nocall` in the output text file. Notice that specifying latitude and longitude causes BirdNET to use a list of 263 species, presumably not including a Swamp Sparrow, as they are unlikely in Colorado.

Swamp Sparrows are common in Florida, and indeed, running the audio file through *analyze fl* produces (256 species):

```
1   Spectrogram   0.0   3.0   Swamp Sparrow   swaspa   0.5129
2   Spectrogram   0.0   3.0   Black-bellied   Plover   0.2532
3   Spectrogram   3.0   6.0   Swamp Sparrow   swaspa   0.9617
4   Spectrogram   6.0   9.0   Swamp Sparrow   swaspa   0.3233
```

with additional detections for the remainder of the file.

BirdNET returns the expected output when told the recording is from an area where Swamp Sparrows are typical. Running *analyze* produces similar detections (all 6522 species considered). Notice that the second detection is, in this case, an error. There was no plover present. Notice also that the detection certainty is low compared to other detections. Like all deep learning models, BirdNET sometimes makes mistakes.

Spotted Towhees are not found in Florida but, though not common, appear in my area nearly year round. As an exercise, I recommend running the file *spotted_towhee.wav* through BirdNET using the location-specific scripts. You'll find the opposite of the Swamp Sparrow results. eBird observation data confirms showing no reported Spotted Towhee sightings in Florida between 2020 and 2025.

BirdNET-Analyzer makes it easy to examine a multitude of recordings. As an exercise, I suggest making a long recording with a digital recorder or a phone using Merlin (though there is a 10-minute limit) and then letting BirdNET do its thing. You might be surprised by what you hear.

11. Going Further

The previous chapters were an introduction, not an ending. Deep learning is much more than the image-classification-centric core presented here. This brief chapter points the way to additional resources and topics you'll find helpful as you continue your explorations.

11.1 Topics for Further Study

We've necessarily focused on image classification. We're interested in birds, and most investigations of birds concern images. We even transformed audio into sonograms so we could continue in the same vein.

Image classification continues with self-supervised learning from unlabeled data. Labeling data is difficult and time-consuming, as we learned. Self-supervised learning takes unlabeled images and defines a proxy task that does not require manual data labeling. For example, training a CNN to learn the number of 90-degree rotations performed on an image has the side effect of conditioning the lower levels of the model to pay attention to the same kinds of features a standard image classifier learns. Therefore, a self-supervised model becomes a good starting point for fine-tuning or transfer learning.

We explored the considerable power found in CLIP embeddings. Expanding that approach to other vision-language models (often called VLMs) is worthwhile in assessing their applicability and understanding of natural images. CLIP can be fine-tuned to enhance its already impressive performance or adapt it to sonograms, an area where we learned it was not particularly well suited. Additionally, it may be helpful to investigate how full semantic segmentation using (possibly pretrained) U-nets can contribute to bird recognition especially between closely similar species.

Sonograms aid in working with audio but including 1-dimensional CNNs, perhaps in an ensemble with sonogram-based models might lead to increases in accuracy or applicability, especially if the audio is filtered in the time domain before sonogram generation.

More generally, we explored supervised learning, but deep learning is also reinforcement learning (RL) and unsupervised learning. Reinforcement learning is behind AI tools like AlphaGo and other game-playing models. Where such models fit in ornithology is for creative minds to decide.

Unsupervised learning is a bit of a catch-all term. Historically, it referred to clustering techniques for data without labels. Such techniques, like the classic k-means algorithm, have a place in modern AI. The term also applies to self-supervised learning, where even older algorithms, such as k-means, might generate pseudo labels for a self-supervised model to learn. It isn't difficult to imagine how this might be used for bird classification: a large, unlabeled collection of images can be clustered according to some criteria or embeddings from a pretrained model and then used to train or fine-tune a model in a self-supervised setting using the cluster number as the class label. The resulting model then serves as a foundation for downstream tasks, much as in transfer learning.

Continuing your exploration of deep learning will benefit from a few good books.

11.2   Recommended Books

The fact that you are reading these words implies you appreciate the utility of books. The texts listed here are a natural extension of our explored material. Many good books are of necessity not listed here. In no particular order, then, you might consider:

Deep Learning
This book by Goodfellow, Bengio and Courville was the first deep learning book. It is deeply mathematical and should be considered a graduate-level text. If you delve deeply into the subject, it is well worth having, but don't view it as an introduction.
(MIT Press, 2016)

Hands-On Machine Learning with Scikit-Learn, Keras, and TensorFlow
Aurélien Géron's book covers the primary toolkits used in this book to explore neural networks and many other kinds of models to provide readers with a broader introduction to machine learning beyond image classification.
(O'Reilly Media; 3rd edition, 2023)

Deep Learning with Python
This book by François Chollet, the creator of Keras, is similar to Géron's book but covers a different mix of topics. Do be aware of the publication date and that the Keras interface has likely changed somewhat from what the book presents.
(Manning; 2nd edition, 2021)

Neural Networks and Deep Learning
Michael Nielsen's online book has been a staple since 2015, and because it's free, it's worth the time to explore and fill in gaps, especially with foundational topics related to neural networks.
(http://neuralnetworksanddeeplearning.com/)

Finally, at the risk of too much self-promotion, I humbly recommend my other contributions in this area:

Practical Deep Learning: A Python-Based Introduction
To me, this book is a logical next step. The material we quickly reviewed in the first two chapters of this book is presented in more detail. If you intend to improve your expertise in deep learning, say to begin a detailed research project, you need more than what we've explored in this book – *Practical Deep Learning* fills in the gaps.
(No Starch Press; 2nd edition, 2025)

Math for Deep Learning: What You Need to Know to Understand Neural Networks
It is entirely possible to be a successful deep-learning user without delving into the mathematics behind it. However, eventually, if for no other reason than curiosity, you'll want to know something about the mathematics – probability, statistics, linear algebra, and differential calculus. Fortunately, it isn't necessary to understand at the level of a professional mathematician. *Math for Deep Learning* covers the essentials of each area as it applies to deep learning in practice. (No Starch Press, 2021)

How AI Works: From Sorcery to Science
This small volume presents AI at tree level – not a bird's eye view (alas) nor a down-in-the-weeds view. I recommend it here because it is an overview of the historical context and an initial exploration of large language models (LLMs), which were not featured directly in the present book, save in part via CLIP. This book tells you where deep learning/modern AI came from and why now. (No Starch Press, 2023)

11.3 Online Resources and Communities

Deep learning resources online are legion. Kaggle has been previously discussed. Here are few you may find helpful.

Coursera
Coursera is one of the first free online learning platforms. Indeed, machine learning was the first course offered. It remains a key resource and offers a deep learning specialization.
(`https://www.coursera.org/specializations/deep-learning`)

Fast.ai
Fast.ai offers *Practical Deep Learning for Coders* which uses PyTorch, the other deep learning framework in popular use. A second course, *From Deep Learning Foundations to Stable Diffusion* continues the exploration up to generative AI.
(`https://www.fast.ai/`)

Reddit
The `r/learnmachinelearning` subreddit is helpful for current news about deep learning and AI. I might also add other social media sites as possibilities, though I personally have no use for them.
(`https://www.reddit.com/r/learnmachinelearning/`)

As with all online resources, caveat emptor.

11.4 The Future of Birding with AI

Birders have always been among us. And that's a good thing. However, birding prior to AI was somewhat a labor of love in that learning to identify birds required a measure of devotion to the study of guidebooks, drawings, photographs and audio recordings. In time, a dedicated birder would learn the essential characteristics of many species, along with their common vocalizations.

The advent of AI breathed new life into the act of birding by endowing everyone with powerful tools to identify the pictures and recordings they make, typically with our ubiquitous smartphones. The Merlin app discussed in Chapter 10 deserves much of the credit for the massive increase in birding in recent years. With AI, casual birding has become commonplace, bringing the benefits of citizen science as an aid to professionals, giving them access to vast numbers of observations spanning the globe. It is no secret that bird numbers have declined in recent decades, often dramatically. Global observations provide the data professionals need to understand the threats birds face from habitat loss and climate change.

AI enhances the ornithologist's toolkit in many possible ways. AI can be applied to tracking bird migration with automatic ID, relieving researchers (and graduate students) of excessive time manually labeling images and recordings. Such wide-ranging, primarily automated identification contributes to real-time diversity estimation to rapidly recognize and respond to decreases in specific areas. Additionally, AI in the form of predictive modeling gives researchers a powerful tool to (hopefully) get ahead of negative population trends and extinction risks.

When birding, I often encounter people with their phones out taking bird pictures or recording audio to help them identify species in the vicinity. I ask "Merlin?" and the answer is almost always in the affirmative. The U.S. Fish and Wildlife Service estimated some 96 million Americans were birding in 2022, more than double the number in 2016.[1] The COVID-19 pandemic combined with apps like Merlin are reasonable explanations for the dramatic increase. I suspect the same effect occurred throughout much of the world.

The explosive growth in AI-assisted birding is not without possible consequences. Clearly, sensitive birding habitats must be protected from disruption by crowds of birders. As with almost everything in life, balance is the key. I see this risk as minimal, manageable and outweighed by the educational benefits AI birding gives benefits that will ultimately lead to a better appreciation of the natural world's beauty, depth and fragility. There is no Planet B; we must care for all the inhabitants of the one we have.

1   https://www.wsj.com/lifestyle/bird-watching-birder-apps-17e1a734

Glossary

Every field has its jargon, including AI. Use this glossary as a reference. It covers the *emphasized* terms used in the book, along with additional terms you'll eventually encounter as you (hopefully!) continue your exploration of deep learning.

activation: The number produced by a neural network node after summing the weighted inputs, adding any bias term, and passing that result through the activation function.

activation function: The function producing the activation for a node in a neural network. The node's inputs, multiplied by their respective weights, are summed, along with any bias term, as the input to the activation function. Activation functions are nonlinear. Examples include rectified-linear units (ReLU), hyperbolic tangent, and sigmoid.

activation map: The collected activations of all the filters or other actions in a layer of a convolutional neural network. In two dimensions, the activation map is a matrix or tensor representing the layer's output, which becomes the next layer's input.

anomaly detection: The branch of deep learning that attempts to label inference-time inputs as anomalous (or out-of-distribution), i.e., something other than what the model was expecting as input.

architecture: The arrangement of nodes and layers in a neural network. The selected architecture influences the function the model can learn from the training set. If the architecture is too simple (small, lacking capacity) the model will underfit (the high bias condition). In code, the architecture emerges as layers are successively added and combined with earlier layers.

artificial intelligence (AI): The branch of computer science that attempts to model intelligent behavior in software or other medium. Historically split between symbolic AI and connectionism. Machine learning is a subset of AI and deep learning is a subset of machine learning. Machine learning builds models from data as opposed to symbolic AI that seeks to codify intelligence via logical statements, scripts and rules.

AUC: The area under the Receiver Operating Characteristic (ROC) curve. A perfect model's ROC curve has an AUC of 1.0, while a model that guesses randomly (binary outputs) has an AUC of 0.5. Because of this, the AUC is sometimes reported with 0.5 subtracted. Known as *Az* in certain circles (e.g., medicine).

background class (see NOTA): *See none-of-the-above (NOTA).*

backpropagation: The algorithm based on the chain rule for derivatives that determines each weight and bias' contribution to the minibatch loss calculated during the forward pass when training a neural network. It's called backpropagation (or backprop) because it works backward from the loss function to earlier layers of the model. The partial derivatives backpropagation calculates are used by gradient descent to update the weights and biases during the backward pass. Modern deep learning toolkits like TensorFlow automatically arrange for the calculation of partial derivatives via backpropagation when the network is constructed in code.

backward pass: The second part of a minibatch gradient descent step. The forward pass calculates the loss over the minibatch while backpropagation calculates the partial derivatives of each weight and bias contribution before gradient descent updates the weights and biases to complete the backward pass.

batch normalization: A layer in deep neural networks that learns how to calculate a scaled and shifted standardized version of the input to the layer during training to keep the values passing through a model within a reasonable range. Batch normalization generally improves model performance.

batch training: Typically means passing all of a training set through a neural network at once during training and using that average loss to calculate the derivatives for a gradient descent step. More generally, batch training uses multiple samples to estimate the gradient of the loss before updating weights and biases, as opposed to single-sample updates.

biases: A scalar added to the sum of the product of a node's inputs and their respective weights prior to passing the total sum through the activation function, $y = \text{ReLU}(\Sigma_i \, w_i x_i + b)$, for inputs x_i, weights w_i and bias b. A traditional MLP is best implemented via weight matrices and bias vectors where each element of the bias vector is the bias learned for a particular node in that layer.

bounding box: A box placed around a detected object in models producing such output, like YOLO. The model often has two output heads, one leading to a softmax or other measure over the set of known classes and another a regression output producing four values for the bounding box in normalized image coordinates, say (x, y, w, h) for the upper or lower corner, (x, y), and the box width (w) and height (h).

CLIP: Contrastive Language-Image Pretraining (CLIP) is a model from OpenAI trained on a vast corpus of image and text description pairs. The word "contrastive" refers to the type of loss function used to coax the model to produce output embeddings in a space where both an image and the text related to it lead to similar locations, or at least vector directions. Because of this joint training, CLIP is able to interpret the meaning present in text phrases and images so that, in this book, a highly accurate, zero-shot bird image classifier was possible by comparing the directions of the CLIP embeddings of the unknown image and a bird's common name.

Cohen's d: The effect size associated with the difference of the means of two datasets. Hypothesis testing wants to know the likelihood that two datasets are samples from the same parent distribution (the null hypothesis). A parametric test, like a t–test, compares means and standard deviations. Nonparametric tests often use rankings. Regardless, a generated p-value that is deemed "significant" (please do better than < 0.05 for that) can be assessed for its effect size using Cohen's d:

$$d = \frac{\bar{x}_1 - \bar{x}_2}{S_p}, \qquad S_p = \frac{\sqrt{(n_1 - 1) \, S_1^2 + (n_2 - 1) \, S_2^2}}{n_1 + n_2 - 2}$$

where an effect size of about 0.2 is small, 0.5 is medium, and ≥ 0.8 is large. It is possible to have a small p-value but the effect size is also insignificant, in which case the result is likely valid but also likely inconsequential.

Cohen's κ (kappa): A metric comparing the proportion of times raters of a categorical model (like a neural network) agree to the probability of agreement by chance. In practice, a value closer to 1.0 indicates a better-performing model. In machine learning, Cohen's κ often closely tracks the Matthews correlation coefficient (MCC), which I recommend as the better overall metric for classification.

confusion matrix: A matrix where each row indicates a test set's true class label and each column the model's output label. A perfect model will correctly classify each test sample by assigning it to the true class. In this case, the confusion matrix is purely diagonal. Mistakes are counts of cases where the model's label and the true label do not agree. Many model metrics are based on the confusion matrix. A confusion matrix for a binary model is sometimes called a contingency table or contingency matrix.

connectionism: An older term for AI based on the emergent behavior of collections of simpler elements, like the nodes (neurons) of a neural network. Connectionism is a bottom-up approach that models the success of living brains, though not by a strict emulation of biological neural behavior. Connectionism was ignored and arguably actively suppressed in favor of symbolic AI until as late as the early 2000s, though the transition can be viewed as beginning in the late 1980s. The explosion of computing power and training data (i.e., the Internet) combined with algorithmic improvements made the hopes of early 1950s connectionist researchers like Frank Rosenblatt a reality.

convolution: A mathematical operation expressed as an integral in continuous spaces but in practice becomes a straightforward discrete operation in neural networks. A kernel of some size, say 3×3, is moved over a larger input (an image). At each position, the product of the kernel values and the input values covered by the kernel are summed and used as a new output value. Convolutions have edge cases addressed by ignoring them (exact or valid) or by zero-padding to produce an output matrix the same size as the input matrix.

convolutional layer: A layer in a convolutional network that learns a set of filters each consisting of a set of kernels convolved spatially over a 3-dimensional input tensor (H,W,C). One filter kernel is applied per channel (C) and summed across channels. The set of such sums, one per filter, becomes the 3-dimensional output tensor. This operation repeats for each sample in the minibatch as it flows through the layer.

convolutional neural network (CNN): A kind of neural network popularized by LeCun in the late 1990s that learns a set of convolutional filters in certain layers. Convolutional networks respond to local structure in complex inputs, like images, and spearheaded the deep learning revolution beginning in 2012. CNNs remain a primary architecture, especially in computer vision.

cosine distance: A distance metric useful in high-dimensional spaces that measures the angle between two vectors. Points in an n-dimensional space are interpreted as vectors with the tail at the origin. The cosine distance, 1 minus the cosine of the angle between the vectors, is 0 for fully aligned vectors, 1 for orthogonal vectors, and 2 if the vectors point in opposite directions. In this book, we used the cosine distance between image and text CLIP embeddings to build accurate bird image classifiers.

cross-entropy loss: A standard loss function mainly used when training neural networks for classification. The cross-entropy loss treats the softmax output vector as a probability distribution and the true label, often a one-hot vector, like so:

$$L = -\sum_i y_i \log (\hat{y}_i)$$

The sum is over classes (i) with y_i the one-hot label for that class (0 or 1) and $\hat{y}_i$ the corresponding softmax vector value. The smaller the loss, the more aligned the two probability distributions are, implying better performance by the model on that sample. The loss is typically averaged over all minibatch samples.

cross validation (k-fold): Splits a dataset into k equal-sized disjoint subsets before training k instances of the same architecture each using $k - 1$ subsets and successive k-th subset held-back as test data. The mean of performance metrics over the k models is taken as a good measure of how well a model trained on the full dataset might perform. This approach is typically used when the dataset is small and the likelihood of random sampling between train and test might lead to pathological performance.

curse of dimensionality: As the dimensionality of model inputs grows (feature vectors or image size), the amount of training data necessary to reasonably represent the space of possible model inputs grows dramatically faster, to the point where simple models like k-nearest neighbors fail because an exponential increase in training data becomes necessary to adequately represent the high-dimensional space. Deep learning is often less affected by the curse of dimensionality because of the size of the datasets and the characteristics of deep neural networks (learning a hierarchy of features).

data augmentation: Adds new training samples constructed by altering existing samples in a way that produces a plausible version of the class represented by the original sample. For example, an image may be augmented by small rotations and shifts or variations in color or scaling. Data augmentation is a powerful regularizer as it conditions the model to learn the essence of classes as opposed to focusing on the characteristics and details of a particular training set.

data drift: Refers to how inference-time data might change, thereby affecting model performance. The prior class probability is learned implicitly during training. If this prior changes over time during inference, the model will no longer perform with the expected level of accuracy. Or, if the characteristics of the data change over time, the model's performance will change. For instance, an imaging sensor might produce snowy, color-faded, or blurry images, all of which will lead to less accurate classification.

decision threshold: The threshold used to determine either a class label assignment (typical of binary models) or whether to assign a label at all or instead flag the inference-time input as too uncertain for classification.

deep learning (DL): The branch of machine learning, and hence AI, that trains large neural networks with many layers. Most modern AI is deep learning.

dense layer: The name often used in CNNs and other modern neural network architectures for a fully connected layer akin to the layers in a traditional MLP, where each element of an input vector is connected to each node of the layer. Weight matrices often represent dense layers and are responsible for the majority of the parameters in a CNN model.

depthwise convolution: Applies a single kernel ($k \times k$) to each channel of the input tensor. If the input is (H, W, C) the output is (H', W', C) where H' and W' are determined by the kernel's approach to edges (valid or same with zero–padding). Depthwise convolution is often followed by pointwise convolution using a 1×1 kernel to mix across channels. The MobileNet architecture uses this approach to reduce the number of learnable parameters and computational burden at inference time.

dropout: A regularization technique applied at training time to randomly zero the output of a layer's nodes with a certain probability (p) to help the nodes learn features without correlation between them. At inference time, the layer's activations are scaled by $1/(1-p)$ to approximate the magnitude present during training. Dropout is typically applied to fully connected layers.

early stopping: A technique applied during training to detect when the model is well-trained and not beginning to overfit. Early stopping typically monitors the validation set accuracy (or other metric) to determine when training should end as opposed to training for a specified number of epochs before stopping. The fact that the validation set is used this way disqualifies it from characterizing the model. For that, the held-out test set must be used.

effective receptive field: Successive layers in a convolutional neural network, prior to the final dense layers present in most architectures, learn features related to larger and larger parts of the input image. The size of these parts is the effective receptive field, the portion of the input image "seen" by the convolutional kernels in each convolutional layer.

embedding space: Convolutional networks map input images to a new output representation, typically flattened to a vector. The top-level dense layers use this vector to classify the input. The flattened vector exists as a point in a multidimensional space corresponding to a new representation of the input image. A well-trained network maps similar input images to similar points (or directions) in this embedding space.

embedding vector: The vector produced by a neural network prior to classification. The vector is often the flattened output of a series of convolutional and pooling layers. The vector can be thought of as a point in an embedding space where similar input images (classes) will map to similar positions or directions in this space.

ensembling: Combines the output of multiple models to leverage the strengths of each and thereby improve classification. The models operate on the same inputs, but they themselves need not be of similar architecture or kind. For example, combining neural network and random forest models in an ensemble is perfectly fine. Combining the outputs might involve averaging softmax vectors in some manner or voting on assigned class labels, etc.

epoch: During model training, a full pass through the training set so that the available training samples have been examined. For minibatch training of neural networks, the ratio between the number of samples in the training set and the minibatch size determines the number of gradient descent steps taken before the training data cycles (is reused for the next epoch).

F1 score: A classification metric that is the harmonic mean of the precision (positive predictive value) and recall (sensitivity):

$$F1 = \frac{2 \times precision \times recall}{precision + recall}$$

F1 scores closer to 1.0 imply a better-performing model. The fact that the F1 calculation does not consider the true negatives (TN) might limit its usefulness in situations where negative predictions matter, e.g., fraud detection.

features: A term for the elements of a feature vector used as inputs to a model. Note that features may be numeric (as in neural networks) or categorical, if the model works with such data (as in a random forest). Categorical features are often mapped to one-hot vectors for neural network models.

feature vector: A term for the input to a machine learning model. In classical machine learning, the input was often a vector, scaled appropriately so that feature ranges were similar if the model type expected them to be (e.g., neural networks). Some models are agnostic to the scale of different features (e.g., random forests). Finally, images, though 2D or 3D if color, are considered "feature vectors" in that they are the input to models expecting such data.

few-shot learning: A technique where models, often pretrained on other datasets, are able to learn new classifications or associations with only a few examples. For example, we used the CLIP embeddings of a handful of examples for each new class (bird species) to build a successful MLP by leveraging the information packed into CLIP embeddings. For large language models (LLMs), few-shot learning is often implemented via in-context learning where the LLM learns the desired task from only a few examples supplied in the initial prompt.

filter: Convolutional layers learn a specified number of filters where each filter consists of a set of kernels, one for each channel of the input tensor. The activations are summed across channels to produce a single output activation for the filter, with the layer's channel output matching the number of filters learned during training.

fine–tuning: Takes a pretrained model, with certain layer's weights often frozen to preserve their existing values, and continues training it with an (often small) dataset to modify the model for a new task or set of classes. The general hope is that the pretrained model, often a foundation model pretrained on a larger dataset, will be easily adapted to a new task for which the training set might be limited. Compare to transfer learning which uses the pretrained model to generate embedding vectors but does not alter the pretrained model's weights.

flatten layer: A neural network layer that unravels a multidimensional input, often the output of a convolutional or pooling layer, into a single vector. Such layers are typically present in CNNs prior to the top-level dense layers or in producing embeddings for downstream tasks. Flatten layers have no learnable parameters.

forward pass: The portion of neural network training where a minibatch is passed through the model using its current weights and biases. The backward pass uses the average loss over the minibatch via backpropagation and gradient descent to update the weights and biases for the next forward pass. Inference is a forward pass without a corresponding backward pass.

foundation model: A loose term for large neural network models of various kinds that required extensive (and often costly) training to produce. For example, OpenAI's GPT series of models or Anthropic's Claude models are foundation LLMs. CLIP is also a foundation model trained on a vast diversity of image and text pairs to make the model's output readily adaptable to a wide range of downstream tasks without extensive alteration.

fully connected model: Traditional neural networks consist of layers of fully connected nodes where every input from the previous layer is connected to every node of the current layer. Such models are fully connected models.

fully convolutional model: Convolutional neural networks that replace dense layers with carefully defined convolutional layers, often with 1×1 kernels. Such models are able to accept input images of any size. Such models respond as if the version of the model using dense layers is convolved over the input image. Fully convolutional models provide a crude form of detection where the class is determined along with some indication of where the classified object is within the input.

global minimum: The local minimum with the smallest value.

gradient: A multidimensional extension of the notion of the slope of a function at a point. The gradient indicates the direction of the steepest function value increase. Therefore, when training or optimizing, the gradient descent step moves in the direction of the negative gradient.

gradient descent (stochastic): An algorithm for locating minima of functions. If the function is convex, there is only one minimum, the global minimum. Nonconvex functions may possess multiple local minima. Gradient descent, which follows the negative of the function's gradient at each position, can become trapped in local minima. Stochastic gradient descent uses an approximation of the true gradient, such as that found by the average loss over a minibatch. The approximate gradient might help the algorithm avoid local minima, thereby leading to a better minimum, implying a better-trained model. Relying on only the gradient (the first partial derivatives) makes gradient descent a first-order algorithm. Nonconvex functions are often optimized using second derivatives, but such an approach is often not possible because of the computationally taxing calculations necessary in the context of deep learning. Therefore, first-order gradient descent is used in the hope that local minima of the loss function are roughly equivalent and that stochastic estimates of the gradient direction will help avoid ineffective local minima.

hard negatives: Training examples that are similar to the target class but are not members of it. For example, suppose Canada goose is one of the target classes. In that case, the training set will benefit from the addition of images of Cackling geese to help the model learn differences specific to Canada geese.

hidden layer: A neural network layer between the input and output layers. Hidden layers, kind and number, make up the bulk of the architecture of a model. What, precisely, hidden layers learn (if learnable) is often difficult to understand, leading to the characterization of models as "black boxes."

hyperparameters: Parameters used in the training of a model that are not part of the static architecture. For example, minibatch size, number of epochs, learning rate, and even optimizer type can be viewed as hyperparameters. Hyperparameter tuning is an important aspect of successful neural network training.

inference: Uses a trained model to evaluate new inputs – inference is the usual end goal of model construction and is equivalent to the forward pass when training.

input layer: The first layer of a neural network that specifies what the network is to operate on. For traditional networks, the input layer is a feature vector. For CNNs, the input is an image.

kernel: A matrix, often small and square, learned by a convolutional layer filter to operate over a particular channel of an input tensor. More generally, for image processing, a kernel is convolved over an input matrix (image) to produce a desired effect, like edge and orientation detection. Convolutional networks employ kernels to learn characteristic representations of parts of the input (image) enabling the model to produce a new representation that is more readily learned by a machine learning model (often the top-level layers of the CNN).

large language model (LLM): A new class of models employing transformer layers to take a series of input tokens (text) and produce a new output token. This process iterates until an end token is generated to produce output representing the model's response to the initial tokens. LLMs have been enhanced to incorporate multimodel input (images, audio, documents, video, etc.) The surprising emergent abilities of LLMs, especially those in the foundation model category (e.g., GPT-4o and the like), have driven the recent rapid increase in AI and have even caused speculation about the nature of thought, intelligence and consciousness as such models increasingly demonstrate human and even superhuman performance in many areas previously believed to be the purview of humanity.

learning rate: Gradient descent relies on the partial derivative ($\partial L / \partial w$) to update a weight or bias. The learning rate (η) is a scale factor on the partial derivative and corresponds to the step size taken in the direction of the negative gradient of the loss function: $w_{i+1} \leftarrow w_i - \eta(\partial L / \partial w_i)$.

LeNet model: An early CNN for classification of handwritten digits (MNIST). We used a version of this model for several experiments.

local minima: A function may have one or more local minima, meaning places where the function value increases on either side of the minimum. The smallest-valued of the local minima is the global minimum. Training a neural network seeks good minima of the loss function.

loss function: Training a neural network involves minimizing a function representing, in some fashion, the errors of the model on the training data. This is the loss function. For classification, the loss function is often the cross-entropy loss that compares the known class label, treated as a probability distribution over labels, with the softmax produced by the model. Regression tasks often use the mean squared error (MSE). More complex learning scenarios minimize more complex loss functions. Additionally, the loss function might involve terms related to regularizing the model so that it learns better (e.g., L2 regularization that penalizes large weights).

machine learning (ML): The overarching term for models learned from data. In the deep learning sense, it refers to neural networks.

matrix: A rectangular array of numbers. Images are represented as matrices. In computer programming, a matrix is a 2-dimensional array. Neural networks are usually implemented as combinations of matrices and vectors.

Matthews correlation coefficient (MCC): A metric that considers all the elements of the confusion matrix (TP, TN, FP and FN for binary models) and is often regarded as one of the best single-number metrics for evaluating model performance. The closer the MCC is to 1.0, the better. The MCC is particularly helpful when characterizing imbalanced datasets and should be preferred to accuracy in those cases.

mean image subtraction: An early approach to normalizing inputs to CNNs. Usually, [0,1] normalization is sufficient, though for particular use cases, say one where the background is complex but consistent across most inputs to the model, might benefit from it. A mean image is formed from the sum of all training samples, and this image is subtracted from the training data and unknown images at inference time.

mean squared error (MSE): The average squared error between an expected value (like a true label or desired continuous model output value) and the model's current output value. The MSE is often used for regression tasks where the model's goal is to predict a continuous value instead of classification, where the model seeks to place an input into one of a finite number of possible classes. The MSE is strongly affected by large deviations from the expected value because the difference is squared (think outliers). It is differentiable, which makes it easy to use in places where gradients are necessary. The mean absolute deviation (MAD) might be more robust to outliers but more difficult to incorporate into calculus-based optimization frameworks. Note that swarm intelligence and evolutionary algorithms are not gradient-based and can work with any concept of "loss" or "error."

metrics: Values determined by a model's performance on a test set. In machine learning, many helpful metrics are derived from the confusion matrix entries. A well-performing model will score highly on many different metrics, in which case the choice of metric is less important (it's obvious the model is doing well). Metric choice begins to matter when the model's performance is more nuanced. In general, I prefer MCC for classification tasks.

minibatch training: Uses a random subset of the available training data for each gradient descent step. While batch training using the entire dataset to estimate the loss seems preferable, in practice, minibatch training is usually better because the noisy loss estimate leads to movement of the model parameters in a noisy (stochastic) way through the loss landscape, thereby potentially avoiding local minima leading to poor model performance. Minibatches are computationally efficient, and stochastic gradient descent benefits the first-order optimization algorithm for training, an algorithm that, naively, should not work for the nonconvex loss landscapes found in deep learning.

model: A generic term for, in this book, a machine learning algorithm with parameters conditioned by a training dataset to accept an input (feature vector or image) and produce an output (class label or regression value).

multilayer perceptron (MLP): A traditional neural network with a feature vector input and fully connected hidden layers. The "perceptron" portion of the name is a reference to the original Mark I Perceptron developed by Frank Rosenblatt in the 1950s.

nearest neighbor (NN): An extremely simple machine learning algorithm where unknown feature vectors are assigned the class label of the nearest training sample. The definition of "nearest" implies a distance metric, typically the Euclidean distance, assuming feature vectors to be points in an n-dimensional space. However, other distance metrics are possible. A k-NN model calculates the distance to the nearest k training samples and assigns a class label based on voting possibly weighted by the distance to each sample. Nearest neighbor models are trivial to implement, and sometimes quite effective. Performance is generally improved by increasing the number of labeled training samples but at the expense of additional computation necessary to locate the k nearest neighbors.

negative predictive value (NPV): The probability that a binary model is correct when it assigns an input to the negative class (class 0). Compare to the positive predictive value (PPV, precision). Formula: $NPV = TN/(TN + FN)$.

neural network: A machine learning model consisting of an arrangement, typically in layers, of individual nodes (neurons) with weighted inputs and a single output value per node. The weights and node-specific values known as *biases* are determined by an optimization process (training) that uses a loss function (error function) and calculated per-parameter partial derivatives (backpropagation) with gradient descent to condition the model to the data. Neural networks are known to be general-purpose function approximators. Traditional neural networks appeared in the 1950s, even arguably in the

1940s, but languished for decades until the deep learning revolution of the 2010s, leading directly to today's advanced generative AI networks.

neuron: A node in a neural network. The term "neuron" refers to the way the operation of the node crudely mimics a biological neuron. A weighted sum of the inputs, including a bias term, is passed through an activation function to determine the neuron's output value.

node: A neuron in a neural network. The term "node" refers to graphs as discussed in computer science. Neural networks are graph structures. In this book, node is the preferred term to emphasize that while there is a loose similarity to biological neurons, the latter are off until on but nodes produce continuous output based on their input.

none–of–the-above (NOTA): A catch-all class, also known as a background class. Used to group possible inputs to a model that are not of interest (not a target class). Without a NOTA class or per-class decision thresholds, many models will assign every input, regardless of type, to one of the target classes.

normalization: Scaling model inputs so that each feature is in [0, 1]. For images, this typically involves dividing pixel values across channels by 255. Some machine learning models, especially neural networks, work best when the features are all within the same range and close to zero. Compare to standardization.

one-hot encoding: Maps categorical values to binary vectors for models, like neural networks, that assume features are ordinal or ratio–valued. For example, biological sex might be encoded as 0 for male and 1 for female, but there is no ordering implied by the values, so a neural network requires the values to be encoded as two binary features: 1 0 for male and 0 1 for female.

optimization: In machine learning, optimization often refers to the process used to train a neural network. Data generates a loss function value that leads to per-parameter gradients ($\partial L/\partial w$), which are then used to update the model's parameters via gradient descent. Optimization in other contexts typically seeks the smallest such loss (error), but machine learning is usually more interested in generalization to new inputs and actively applies regularization techniques to prevent the optimization process from focusing on specific characteristics of the training set in favor of the general characters leading to robust discrimination.

optimizer: The specific optimization algorithm employed when training a neural network (in the context of this book). The simplest optimizer is stochastic gradient descent (SGD), which uses the loss over minibatches to estimate the loss gradient and therefore the step direction for parameter updates. More advanced optimizers are based on gradient descent but include extra calculations to better guide the learning process. Adam is one such advanced optimization algorithm (used frequently in the book's experiments).

out of distribution (OOD): A generic term for model inputs outside of the kind used to train the model. Detecting and responding to OOD inputs is an active research area. An OOD input might be considered an anomaly, depending on context.

overfitting (high variance): Describes a model that has learned minute details or characteristics of a dataset instead of general characteristics that lead to good performance in the wild. Also referred to as high variance, an overfitting model fails to perform well at inference time because it has learned the training set so well that it cannot adequately capture what characterizes an arbitrary input. A good analogy is fitting a high-dimensional polynomial to a dataset so that the error over the measured points is near zero, but the function varies wildly at other points.

parameter: A generic term for a value in a machine learning model conditioned by training data. For neural networks, weights and biases are parameters, and parameters of the training process are called hyperparameters.

Perceptron: The Mark I Perceptron developed by Frank Rosenblatt in the late 1950s is often considered the original neural network device. It was capable of simple classification, even of crudely digitized images, but lacked the flexibility necessary to capture complex functions mapping inputs to outputs. As such, neural networks were (incorrectly) perceived by many computer scientists to be limited and unlikely to be an essential component of artificial intelligence. The Perceptron was a connectionist device, and because of its limitations, symbolic AI dominated artificial intelligence research until the 1980s and the deep learning revolution of the 2010s. Rosenblatt's enthusiastic claims for the Perceptron's eventual abilities aided skepticism of the approach, though it is worth noting that most of those claims were eventually realized with modern neural networks.

pooling layer: A layer commonly used in convolutional neural networks to reduce the spatial dimensions of a tensor while preserving the number of channels. A pooling kernel, often 2×2 with a stride of 2, moves over each channel selecting a single value in the kernel, often the maximum value. Pooling reduces the number of parameters in a tensor at the expense of spatial information contained in the tensor. Advanced networks increasingly use strided convolutions or attention mechanisms (transformers) instead of pooling.

positive predictive value (PPV): The probability that a binary model is correct when it assigns an input to the positive class (class 1, target class). Also known as *precision*. Compare to the negative predictive value (NPV). Formula: $PPV = TP/(TP + FP)$.

prior probability: The probability with which a class appears naturally in the wild. Models implicitly learn prior class probabilities from the training data, and it is important to preserve model performance that these prior probabilities are consistent when the model is used for inference. More generally, a prior probability appears in Bayesian statistics as the natural probability with which something appears (e.g., frequency of four-leaf clovers) or still more generally as an intrinsic belief in the occurrence of an event.

Receiver Operating Characteristic (ROC) curve: A graph parametrically generated, for a binary model, by varying the decision threshold over $[0, 1]$ and then using the resulting confusion matrices to calculate the true positive rate (sensitivity, recall) and the false positive rate (1 − specificity) for each threshold. The curve formed by plotting (FPR, TPR) pairs is the ROC curve. The elbow of the ROC curve moves toward the upper left corner as the model's performance increases. The area under the curve (AUC) is sometimes used as a single-number metric with 1.0 the maximum. The term ROC originated during WWII among early radar engineers to assess signal detection performance.

regression: A machine learning (and curve fitting) technique that generates a continuous output value instead of a class label. For example, a regression model might accept an input feature vector describing the characteristics of a house and output a price for that house. Regression in curve fitting assumes a form for the model and learns the parameter values, while machine learning assumes nothing about the functional form.

regularization: Anything that reduces a model's tendency to overfit and instead learn general features of the training set. Regularization techniques include gathering more training data, data augmentation, simulated data, and terms applied to the loss function (L2 regularization for weight decay) to nudge the network's weights and biases to values leading to better generalization. Additional regularization approaches include batch normalization and dropout layers or early stopping based on validation set metrics.

ReLU: A rectified linear unit activation function. ReLU is, in its most basic form, nothing more than $\max(0, x)$ to return the input, x, if $x > 0$, otherwise return 0. Leaky ReLUs relax this restriction to allow negative values through according to a linear mapping with a small slope, often learned during training. Modern networks use ReLU activation functions extensively because they typically perform better than more traditional activation functions and are trivial to implement. Consider the number of terms necessary to implement a Taylor series approximation of an exponential function (needed for hyperbolic tangent and sigmoid activation functions) to the single line of code to implement a ReLU.

self-supervised learning: An approach to learning with unlabeled data where a proxy task is defined, such as learning which 90–degree rotation was applied to an image, to create pseudo labels for model training. The trained self-supervised model is then used downstream for transfer learning or fine–tuning.

semantic segmentation: A detection technique that assigns a class label to every pixel of an input image. Semantic segmentation uses models like U-Net to output a set of class labels with the same spatial dimensions as the input image. In semantic segmentation, all instances of a particular class receive the same label. Instance segmentation goes one step further to identify each instance of a class, along with all of its pixels.

sigmoid: Once used as an activation function (upstaged by ReLU), sigmoid functions, also known as logistic functions, can be used in place of a softmax for binary models where class 1 is the target and the network's output is a measure of the model's confidence that the input is a member of the target class. Mathematically, $\sigma(x) = 1/(1 + e^{-x})$, which is defined for all x forming an S-shaped curve between 0 and 1 with value 0.5 at $x = 0$.

softmax: A multiclass extension of the sigmoid to map multiple model outputs to a pseudo-probability distribution so that the sum across the elements of the softmax vector is 1. Classification usually assigns a class label based on the maximum softmax vector value.

standardization: A technique for mapping features to zero mean and standard deviation one:

$$x' = \frac{x - \bar{x}}{\sigma}$$

Many machine learning models expect the features to be in a common range about a value near zero. Compare to normalization.

symbolic AI: An approach to artificial intelligence that relies on symbol manipulation and logical statements or rules. Symbolic AI dominated connectionist approaches to artificial intelligence research for decades until the deep learning revolution of the 2010s. Symbolic AI is still being explored and often used with neural networks.

tensor: In a machine learning context, a tensor is a multidimensional array representing a dataset. Deep learning toolkits like TensorFlow and Keras work often with 4–dimensional tensors, (N,H,W,C), for N samples each of height H, width W and number of channels C. For example, a minibatch of 64 32×32 pixel RGB inputs to a model can be represented as a (64, 32, 32, 3) tensor. A scalar is a 0D tensor, a vector a 1D tensor, a matrix a 2D tensor, and a volume a 3D tensor.

test set: In supervised learning, a collection of labeled samples held back during model training for calculating final metrics after training is declared complete. The test set should match the general characteristics of the training set.

training set: The collection of labeled samples used to train a model. The richer the training set in terms of quantity of variation and representativeness of what the model will encounter at inference time, the more likely it is that training will be successful and the model suitable for the intended purpose.

transfer learning: Uses a pretrained model to generate output embedding vectors to map a dataset to a new representation. The mapped dataset is then used for downstream tasks, such as training a simpler model. Transfer learning helps in building models for new classes or with limited training data by leveraging what the (often larger) pretrained network has learned.

true negative rate (TNR): The true negative rate (specificity) is the probability that a true negative sample will be classified as a true negative by the model: $TNR = TN/(TN + FP)$. Compare to the negative predictive value (NPV).

true positive rate (TPR): The true positive rate (sensitivity, recall) is the probability that a true positive sample will be classified as a true positive by the model: $TPR = TP / (TP + FN)$. Compare to the positive predictive value (PPV, precision).

t-SNEplot: t-distributed stochastic neighbor embedding is a dimensionality reduction technique useful for visualizing high-dimensional data in two or three dimensions by attempting to preserve local relationships present in the high-dimensional space. A t-SNE plot displays such mapped vectors to aid in understanding the data.

uncorrelated: Refers to features that do not change in a related way where one increasing also implies the other is increasing (or decreasing). Correlated features are difficult for machine learning models to work with. Image pixels are usually highly correlated (neighboring pixels are very much like each other). Convolutional neural networks cope with such correlations by learning a new representation of the input image, one that a traditional neural network (the top layers) can learn more successfully.

underfitting (high bias): If a machine learning model lacks the capacity to learn the true function relating training inputs to outputs, the model will underfit and not capture essential characteristics that allow it to generalize well to unknown inputs. In other words, the model will exhibit a high bias toward the simpler model. For example, a line will underfit a quadratic function and not perform well at interpolating new function values. The analogy holds with low-capacity machine learning models.

validation set: In supervised learning, a validation set is a third dataset after the training and test sets used during training to assess the training process. For example, the validation set's performance might be used to implement early stopping before overfitting or to determine that the model architecture needs to change. Because of this, it is philosophically disingenuous to report validation set performance as an assessment of the model. A held-out test set must be used.

vector: A one-dimensional (1D) tensor is usually implemented in code as an array with a single dimension. Mathematically, vectors can be thought of as $n \times 1$ (column) or $1 \times n$ (row) matrices. Vector and matrix operations are at the heart of deep learning and efficiently implemented on modern GPUs.

weight decay (L2 regularization): A regularization technique that adds a loss function term to penalize large network weights. In this form, weight decay is called L2 regularization. The same effect can be achieved by adding a term to each weight's gradient descent update equation.

weights: The values multiplying the inputs to a node in a neural network. Weights and biases (node–specific) form the parameters of the network. Training uses data to estimate these values.

zero–padding: Exact convolution drops values at the edge of the matrix where the kernel attempts to multiply values that don't exist. Zero-padding imagines zeros for these values so that the output matrix is the same size (number of rows and columns) as the original. Toolkits indicate zero-padding should be used by selecting "same" for the convolutional layer mode.

zero-shot learning: Using a model as-is for a new task without altering the model by giving it additional examples to learn from. In this book, we used CLIP out of the box to build accurate bird image classifiers without fine-tuning or altering the model (when using cosine distance as the metric).

Index

www.ingramcontent.com/pod-product-compliance
Lightning Source LLC
Chambersburg PA
CBHW061926190326
41458CB00009B/2668